The Botanist and the Vintner

By the same author

THE MAHARAJAH'S BOX:
*An Imperial Story of Conspiracy,
Love and a Guru's Prophecy*

FENIAN FIRE:
*The British Government Plot
to Assassinate Queen Victoria*

The Botanist
and the Vintner

HOW WINE WAS SAVED
FOR THE WORLD

CHRISTY CAMPBELL

Algonquin Books of Chapel Hill   2005

Published by
Algonquin Books of Chapel Hill
Post Office Box 2225
Chapel Hill, North Carolina 27515–2225

a division of
Workman Publishing
708 Broadway
New York, New York 10003

First published as *Phylloxera: How Wine Was Saved for the World* by
HarperCollins Publishers, London, in 2004. Published by Algonquin Books
of Chapel Hill in 2005.

Printed in the United States of America.
Published simultaneously in Canada by Thomas Allen & Son Limited.

Library of Congress Cataloging-in-Publication Data

Campbell, Christopher, 1951–
    Phylloxera
    The botanist and the vintner : how wine was saved for the world /
Christy Campbell.
        p. cm.
    Originally published: Phylloxera. London : HarperCollins Publishers,
2004.
    Includes bibliographical references (p.).
    ISBN-13: 978-1-56512-460-8
    ISBN-10: 1-56512-460-X
    1. Grapes—Disease and pests—France—History—19th century. 2.
Viticulture—France—History—19th century. I. Title.

SB608.G7C3199 2005
634.8′27′094409034—dc22

                                                    2004059781

10 9 8 7 6 5 4 3 2 1
First Edition

For Clare

CONTENTS

MAPS

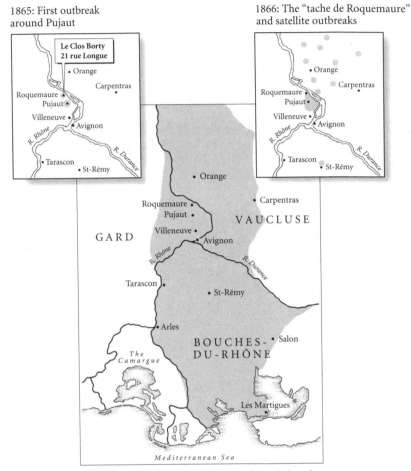

1865: First outbreak
around Pujaut

Le Clos Borty
21 rue Longue
• Orange
Carpentras
Roquemaure •
Pujaut •
Villeneuve •
• Avignon
R. Rhône
R. Durance
• Tarascon
• St-Rémy

1866: The "tache de Roquemaure"
and satellite outbreaks

• Orange
Carpentras
Roquemaure •
Pujaut •
Villeneuve •
• Avignon
R. Rhône
R. Durance
• Tarascon
• St-Rémy

• Orange
Roquemaure •
Pujaut •
Villeneuve •
• Avignon
R. Rhône
• Carpentras
VAUCLUSE
GARD
R. Durance
Tarascon •
• St-Rémy
• Arles
BOUCHES-
DU-RHÔNE
• Salon
The
Camargue
Les Martigues
Mediterranean Sea

By 1867–68 the whole lower Rhône valley is infected

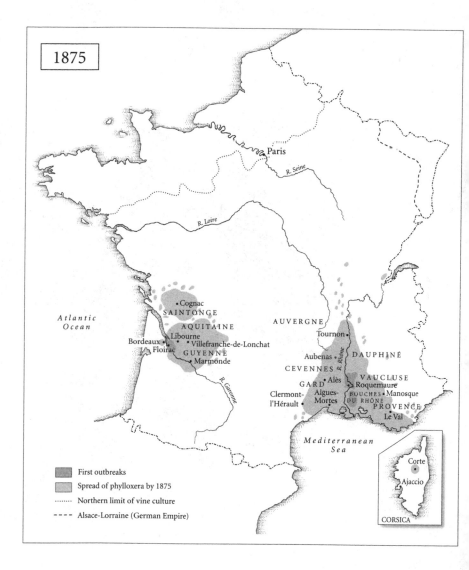

1875

Paris

R. Seine

R. Loire

Atlantic
Ocean

Cognac
SAINTONGE

AQUITAINE   AUVERGNE
Libourne   Tournon
Bordeaux • Villefranche-de-Lonchat
Floirac   Aubenas •   DAUPHINÉ
GUYENNE   CEVENNES
Marmonde   VAUCLUSE
GARD • Alès   Roquemaure
Clermont-   Aigues- BOUCHES • Manosque
l'Hérault •   Mortes   DU RHÔNE
PROVENCE
Le Val
R. Garonne

R. Rhône

Mediterranean
Sea

Corte

Ajaccio

CORSICA

First outbreaks

Spread of phylloxera by 1875

Northern limit of vine culture

Alsace-Lorraine (German Empire)

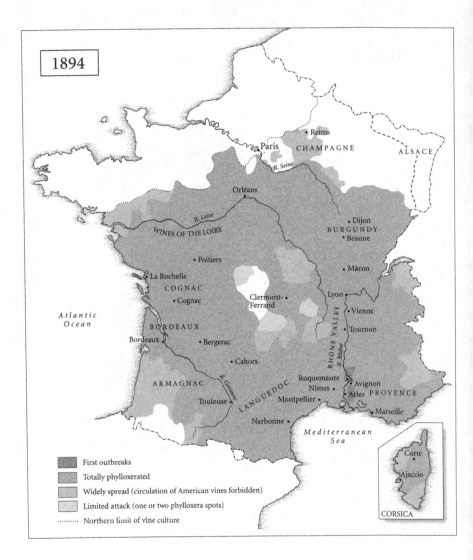

1894

Reims
Paris • CHAMPAGNE
R. Seine
ALSACE

Orléans

R. Loire
WINES OF THE LOIRE
• Dijon
BURGUNDY
• Beaune

• Poitiers
• Mâcon

La Rochelle
COGNAC
• Cognac
Clermont- •
Ferrand
Lyon •
• Vienne

Atlantic
Ocean
BORDEAUX
• Tournon
Bordeaux •
• Bergerac

RHONE VALLEY
R. Rhône

• Cahors

ARMAGNAC
R. Garonne
Roquemaure •
Nîmes •
• Avignon
Arles •
PROVENCE
Toulouse •
LANGUEDOC
Montpellier •
Marseille •

Narbonne •

Mediterranean
Sea

CORSICA
Corte •
Ajaccio •

First outbreaks
Totally phylloxerated
Widely spread (circulation of American vines forbidden)
Limited attack (one or two phylloxera spots)
.... Northern limit of vine culture

GLOSSARY

alate – (of insects) winged

apterous – wingless

baptistère – "baptizer" of phony vintages, a fraudster

betteravier – sugar-beet grower

cépage – French term for vine variety (or vine species)

clos – walled vineyard

coupage – cutting and blending

dégustation à l'aveugle – blind tasting

encépagement – matching vine variety or species to local soil and climate conditions

en cuve – in the fermenting vat

franc de pied – vine growing on its own roots

fuchsine – red dyestuff (containing traces of arsenic) used to color sugar or raisin "wines"

grafted vine – plant artificially spliced together to make not a hybrid but a functioning compound organism from genetically separate components

grand cru – term originally used for the highest quality wines of Bordeaux and later those of Alsace, Burgundy and Champagne

greffon – graft scion

hectare – a metric unit of measure equivalent to 2.471 acres. Wineries in Europe use this term to describe the land area of vineyards. Output of wine is measured in hectoliters per acre. A hectoliter is equivalent to 100 liters or 26.4 U.S. gallons hybrid – interspecific plant originating in the wild or artificially cross-pollinated from two or more species

mas – farm; Provençal equivalent of *château* or *domaine*

négociant – wine merchant

pal – outsize syringe used to inject insecticidal chemicals into the ground

pépinière – nursery

producteur directe – non-grafted American vine species or hybrid used for direct production of wine grapes

rootstock – below-ground section of grafted vine, typically derived from a "phylloxera-resistant" American vine species or hybrid

terroir – the sum of a vineyard's elements – especially soil, but also topography and climate

variety (or "varietal") – a vine with distinct characteristics, a man-made creation from a vine species, typically cloned from generation to generation by asexual propagation. Merlot, Gamay, etc. are varieties of Vitis vinifera; Clinton, Jacquez, etc. are varieties of American vine species or interspecific hybrids

vendange – grape harvest

vigneron – vineyard worker

vignoble – vineyard (can refer to that of the entire country – e.g., *le vignoble français*)

vinifera – shorthand for Vitis vinifera, the "European" vine species from which all classic wine-producing varieties are derived

"It is called wine," said O'Brien with a faint smile. "You will have read about it in books, no doubt."

George Orwell, *Nineteen Eighty-Four*

PROLOGUE

California, August 1994

The white-painted aircraft turned in lazy circles, its sailplane wings scrabbling at the thin air of the upper atmosphere. California lay spread out twelve miles below. As the pilot took course northwards, bouncing gently on the thermals, the peaks of the Sierra Nevada shimmered in the heat haze marking the starboard horizon.

Designed in the depths of the Cold War to sniff out Soviet missile silos, the sensor-laden aircraft had been adapted to the needs of a new owner and a new mission. It was still looking for an enemy. From the edge of space, it was in search of an insect.

The NASA pilot tracked the line of Highway 29, running in a shallow arc along the base of the Napa Valley from San Pablo Bay to the mountains above Calistoga, its flanks corrugated green by marching rows of vines. They were methodically set ten meters apart, enough for a tractor to pass between them. This man-made grid was useful. The Lockheed ER-2's navigational computer took fixes from orbiting navigation satellites and transponders on the ground. The infra-red film camera stared down, each high-speed motor-driven frame automatically collated with a precise ground-plot. The technology was sophisticated but the scientific premise very simple. The summer vineyards below showed fine variations in temperature according to the mass of their foliage, the "reflectance" of their leaves and chlorophyll density.

Anything causing "stress" to the vines – drought, anomalous soil conditions, attack by parasitic pathogens or insects – would be picked up by the infra-red and show up in computer analysis. On the resulting images red indicated bare soil, "through tan, yellow, light green, and deep green, corresponding to increasing canopy cover and plant health."

At this altitude, the whole of Napa and Sonoma counties could be covered in under an hour. But the area of special interest spanned seven hundred acres, at the northern rim of the Mondavi vineyards around the little town of Oakville. An old enemy was there.

Oblivious to the unblinking airborne eye sixty-five thousand feet above them, tourists sheltered from the midday heat sipping *fumé blanc* in the agreeable tasting room of the Robert Mondavi Winery, flamboyantly fashioned in the Spanish mission style. Founded three decades before by the son of Italian immigrants, it was the proudest citadel of California's golden new industry.

Napa Valley had become its billion-dollar heart. Corporations (Coca-Cola had bought in, then bundled out) had raised ever more baroque wineries – Rhenish castles, Greek monasteries, Médocain white-stone châteaux. Since the 1960s, burned-out bankers, pro golfers and retired Berkeley professors had found footholds in the new Eden. The product got better and better. Land prices soared. Then the enemy had returned.

It was a tiny insect, an aphid, that in its native eastern and southern North America lives parasitically on certain species of one genus of plant – Vitis, the grape-vine.

The insect had been called many things since it was first identified by a New York entomologist 130 years earlier. It reached the height of its infamy when it reigned over the vineyards of Europe and California in the last three decades of the

nineteenth century. Then it had been called *Phylloxera vastatrix*.
Modern science knows it as *Daktulosphaira vitifoliae*.

In 1983 the owners of a small vineyard at St. Helena, a few
miles north of Oakville, had reported a cluster of vines failing
to thrive. Their leaves had bunched into ugly "cabbage-heads."
A scientist from the University of California arrived to investigate, suspecting nematode (wire-worm) infection as the cause.
The vines were dug up and their roots examined. They were
encrusted with tiny yellowy-orange aphids. This was not meant
to happen.

The infestation spread. In July 1992 the *San Francisco
Chronicle* reported:

> *The insect plague known as phylloxera has gone from
> isolated cases in a few vineyards to a full-fledged infesta-
> tion. Experts say some fifty thousand acres in Sonoma and
> Napa counties will have to be bulldozed and replanted
> with phylloxera-resistant rootstock in the next ten years.
> The cost of replanting, including loss of revenue while
> growers wait three years for new vines to come into pro-
> duction, is estimated at $500 million to $1 billion . . .*

A century before, after years of argument and denial, an extraordinary scientific effort in Europe and the United States
had begun to roll back the phylloxera's predations. It had been
discovered, in a lesson in the then still very controversial theory
of natural selection, that the aphid and its various American
host species had found biological mechanisms for mutual coex-
istence. The single species of European vine, Vitis vinifera, cultivated for thousands of years by man to produce a myriad of
wine-yielding varieties, had no such defense. When living
aphids were accidentally transported across the Atlantic in the
early 1860s, they fell on the helpless Europeans, sucking and
corrupting their roots until it seemed there would be none left
to consume.

Primitive insecticides proved expensive and dangerous. The solution came from the vine itself: European stems were grafted to American roots to create not a hybrid but a binary compound organism, conjoined and functioning in its structure but genetically quite separate.

For more than thirty years the vineyards of France had acted as a vast laboratory in which experimenters labored to find "rootstocks" that worked. They had to be effective in varying soil types, and successfully take grafts of vinifera. Researchers established that within American native species and cultivated sub-varieties and hybrids* of those species, there was a wide range of resistance that could be experimentally rated. It was further discovered that American species could be directly crossed with Old World vinifera to make genetically hybrid rootstocks with subtle advantages in vigor, tolerance to lime, drought and salinity, and from which more such vines could be successfully propagated.

Grafting, eventually, had worked in France – and it had worked in California, where an industry as old as the first Spanish settlements had originally been built on European vines. That too had succumbed when, in the 1870s, the aphid had crossed the Rocky Mountains. Like the vineyards of Europe, those of Napa and Sonoma had long ago been uprooted and replanted with grafts on a rootstock that science had proclaimed would defeat the insect's attack. Now, more than a century later, the defense had begun to fail.

The search for an answer went into the upper atmosphere. In the early 1990s scientists from the University of California and

---

* Members of the genus Vitis break the classic definition of a species in that interspecific hybrid crosses between them are fertile, and remain so down succeeding generations. Vitis has been described as a genus of "ecospecies," their differentiation produced by adaptation to singular environmental conditions.

the Robert Mondavi company began working with NASA's Office of Advanced Concepts and Technology, which provided the aircraft, remote sensing and computer imaging. The program was called GRAPES – the Grapevine Remote-sensing Analysis of Phylloxera Early Stress. For three summer seasons the Oakville vineyard and those beyond were scrutinized, the remotely sensed data feeding into a computer matrix at NASA's Ames Research Center at Moffett Field, in the heart of California's Silicon Valley.

On the computer imagery, rivers of pale-green, yellow and tan were spilling outwards, showing the presence of the insect underground.

The phylloxera-hunters of the 1870s, with their hand lenses and collecting jars, could never have dreamed of such research tools. But they would instantly have recognized the findings – the aphid was spreading remorselessly, apparently travelling on farm machinery, in the subterranean drainage system, and with irrigation and rainwater run-offs.

"There is no way to eradicate the pest," said the University of California report delivered in 1996; "the infestation is present in eight California counties, and is particularly severe in Napa and Sonoma Counties where thousands of acres of premier vineyards have already been destroyed . . ."

A century ago science had gotten many things wrong in the battle against the phylloxera. But botanists and entomologists had eventually gotten several fundamental things right in their understanding of how the genus Vitis, in its many species and varieties, resisted the aphid's attacks. Now, at the outset of the third millennium, it seemed that accepted science was flawed. The vine was just the same. It was, said the California researchers, the insect that was different.

Part One

DENIAL

1

Paris, October 1867

Mr. Marshall Pinckney Wilder sniffed the cork contemplatively. The wine-waiter's eyes flickered with alarm. The gentleman from Massachusetts and his two companions had already navigated the *carte de vin*, itself a novelty with the new fashion for estate-bottled wines, with all the expertise of English *milords*.

The Exposition Universelle, the great exhibition of the world's arts and manufactures, had packed the hotels and restaurants of the imperial capital with visitors all summer. But so far few transatlantic patrons of the Café Anglais had known the difference between a Bordeaux and a Burgundy. Certainly not in this establishment, but less scrupulous restaurateurs knew that Americans would drink anything that was put in front of them.

Now here were three of them, in felt hats and fancy waistcoats, commenting knowledgeably on the classification of the Bordeaux *crus*, on intricate variations in the soil of the Côte d'Or, on the merits or otherwise of crushing grapes with bare feet. As they celebrated the end of their viticultural fact-finding mission, the little party discoursed on the latest scientific theories of M. Pasteur on why grape-juice fermented into alcohol in the first place. It was something to do with "bacteria."

The task given them that spring by the U.S. government had seemed as pleasant as it was forward-looking. Other officials had been dispatched by Washington to examine the latest advances in ammonia production or railroad engineering as the nation strove to rebuild after the Civil War. Wilder and his colleagues had been sent to drink wine. Why should the United States not have a great national wine industry, and thus release its citizens from the grip of the beer barons and hard liquor?

Good wine was a blessing. "When Americans consult French physicians, three times in four they are ordered to drink red wine as a habitual beverage," Wilder would report. "The greatest longevity is among those people who take red wine three times a day and abstain from both tea and coffee." The trouble was getting hold of the right stuff.

Thus Wilder, strawberry-breeder and Bostonian wine connoisseur, had been appointed chairman of the committee sent to France both to represent America's tiny wine industry and to learn what they might from the world's greatest. As it turned out, neither task had been problem-free.

The Massachusetts fruit farmer (Wilder had founded the American Pomological Society) had seen trouble ahead when not a single American was appointed to the Exposition's judging panels who would rule on the quality of wines from all over the world. "It was to remedy this omission that a special committee was appointed, to examine the wines of our own and other countries, and report especially with reference to vine-growing in America," wrote Wilder. He had to admit their palates were not the most refined. His fellow commissioner, the Ulster-born Patrick Barry, noted nurseryman of Rochester, New York, knew a lot about fruit trees but nothing about wine. William J. Flagg, however, amateur architect and landscape gardener of Cincinnati, thought himself quite an expert.

The commissioners had spent the three months beforehand poking into the crevices of the Médoc, Burgundy and Champagne, seeing how the French grew grapes and made wine. They had come away admitting utter bafflement. But at least they had learned something about the finished product along the way. Wilder had declared the party's choice from the cellar of the Café Anglais with confidence.

A '65 from Messrs. Averou Frères of Pauillac, an obscure, unrated *cru* but an excellent decision nevertheless. At this particular vineyard, so Wilder reminded his companions, no peasant feet were employed in tramping out the vintage. They knew – they had been to the commune of Haut-Bages to inspect the modern "grape-mill" ready for the coming autumn harvest. Little by little such new scientific practices were beginning, in some pockets of enlightenment at least, to transform an industry far older than France itself.

Marshall P. Wilder swirled the dark red liquid, sipped delicately and dabbed his enormous moustache with a linen napkin. There was a grunt of approval.

Their task had not always been so agreeable. Being an American in Paris could mean being played for a sucker. "Reliance cannot be placed on what is furnished to the traveller at hotels and cafés no matter what price he pays," Wilder would comment in his report. "It is not unusual for dealers to buy of producers in the back country a coarse red wine for thirty cents a gallon, mix it with white wine and bottle them and send them abroad with all the high sounding names of the saints of Médoc to sell at enormous profit to unsuspecting foreigners."

As they had travelled that summer by steamer down the Gironde River northwards towards Pauillac, the party observed cargo ships of every nationality heading for the great wine port of Bordeaux. "One of them bore the flag of my country, and as I gazed on its folds I knew it would soon be waving proudly over a homeward-bound cargo of as inferior liquor as Bordeaux could export," noted Flagg.

The Anglo-Saxon press had been full of such horror stories ever since the attack a decade before of the *oïdium* – a fungal parasite which had appeared seemingly from nowhere, and seemed set to wipe out much of southern Europe's vineyards. French wine production had collapsed. *The Times* had reported indignantly how Spanish wine was being turned into "Bordeaux" by adding plaster and orris-root. A plethora of such scandalous "receipts" tended to appear in the London press every time there was an Anglo-French war scare, which was not infrequently. But a remedy to the blight had been found, using powdered sulphur as a fungicide.

By the time the model vineyard plots were being laid out on an island in the river Seine at Billancourt, the agricultural outstation of the Exposition, French wine-makers had regained both their global predominance and some of their old arrogance.

There was the matter of the blind wine-tastings (*dégustations à l'aveugle*) held in a specially-built rustic pavilion on the little island. At every session French judges gave straight zeros to "Catawba" wine, proud product of Ohio, made from grapes borne by a native American vine. "The more of the natural flavor the wine possessed . . . the lower they would estimate it," the judges had loftily informed Wilder. "It was rather too much to ask of the French members especially, to fall in love with what seemed to them new and *sauvage* aromas and flavors," reported Flagg.

The German jurors had shown no such prejudice. Indeed they had gone out of their way to commend the sparkling beverage made from the Catawba grape by Messrs. Werk & Sons, late of Alsace, now of Cincinnati, Ohio.

The best America could do was an honorable mention for "champagne" made by the Buena Vista Viticultural Society of Sonoma, California. That itself was a backhanded compliment. The jury was commending an American copy of a European original. Everyone knew that west of the Rocky Mountains, Old World vines, originally imported by the Spanish, had prospered. But anywhere else in North America, the European grape-vine species, Vitis vinifera, had shrivelled and died within a few years of transplanting. No one knew why. The soil itself seemed poisoned. But the vast continent abounded with wild species, several of which had been domesticated over many years into wine-yielding varieties such as Catawba by diligent French and German immigrants.

Why, Catawba was famous back home. The poet Longfellow had written some very bad verse in its honor. According to legend, in 1818 Major John Adlum, veteran of the War of Independence and amateur vine-grower, had "found" the native species Vitis labrusca, domesticated into a wine-yielding variety by a German family in a garden near Washington, D.C.

The old soldier's vine had been taken up by an energetic Cincinnati lawyer-turned-entrepreneur named Nicholas Longworth who in the 1840s had pursued his vision of turning the banks of the Ohio river into the "Rhine of America." He had fortunately abandoned the major's practice of adding pounds of sugar to the "must," or making wine from wild grapes when his own harvest was insufficient, and had produced something much more palatable. In a bid to make a sparkling wine, Longworth had recruited a French cooper from Reims, but he had unfortunately "drowned in the Ohio a few days after his arrival."

Nicholas Longworth's efforts had prospered, however. Over four thousand hectares of Ohio vines were under cultivation by the time of the Exposition, tended by an army of German immigrants. The American Wine Company of St. Louis, Missouri, even re-fermented Ohio-produced Catawba into something they called "champagne." But the tutored palates of European visitors who tried such beverages could never forgive any product of the labrusca grape's curious animally undertaste. "Musky," they called it – "foxy" – *le goût de renard*. The polite called the taste *framboise*, "raspberry." The more direct called it *pissat de renard* – "fox piss."

There were, it had to be said, some Frenchmen in France itself who had been experimenting for many years with vines of American origin. Scientifically minded *viticulteurs* (and there were a growing number) had been importing samples of New

World vines, with their strange frontier names, since the end of the Napoleonic wars. For some unknown reason they had proved immune to the wholesale attack in the 1850s of the *oïdium* – when European vines had succumbed wholesale to the hateful fungus.

"A French wine-grower has introduced the Catawba into his vineyard," Wilder reported, "and uses its juice to mix in very small proportions with those of native grapes." But, he felt compelled to add, "any considerable addition of the Catawba's musky quality would be more than the French palate, trained to like all that is negative, could very well bear."

That was the problem. The gulf in transatlantic wine culture was too wide ever to be bridged. The French were too exquisite or too devious, Yankees too dumb. "American consumers have palates as yet so unskilled, and the merchants of Bordeaux – the fabricators and imitators – are so adroit, it seems impossible for the honest wine-grower here to come into such relations with [our] wine-growers as will secure to the latter the benefits which the French people themselves derive from the pure juice of the grape so abundantly produced in this country," Wilder reflected sadly from autumnal Paris.

And what lessons might they themselves bring back from the universal exhibition? None, it seemed. The committee's report ended: "There seems to be no one method in use here, in any stage of vine-raising or wine-making. There is such a confusion of practice and a conflict of theory such as it would be hopeless to attempt to reconcile."

Dr. Jules Guyot, the French imperial government's very own vine expert, could not but agree. "Viticulture in France is in the grip of complete anarchy," he wrote in his own official report on the state of vine-growing and wine-making as demonstrated in Paris that late summer. "An infinite number of practices, including the strangest and the most contradictory, are applied to the vineyard without any principle or rules . . .

"Each wine-making province, each *département*, each canton is convinced that its traditional methods are the last word in the art and the science of wine. Every wine grower is persuaded that no one else can make wine better than him. The moment has come to at last bring these practices into the daylight, compare like with like and test them by open competition. This initiative will be one of the glories of the Exposition Universelle of 1867."

Dr. Guyot could not have imagined the scale of the terrible test that was about to engulf France.

2

Powerscourt Demesne, Co. Wicklow, Ireland, summer 1867

Cultivating the vine in Ireland might have been thought an eccentric activity. But growing grapes, "the most noble and challenging of fruits" in the words of Malcolm Dunn, head gardener to the Seventh Viscount Powerscourt, was an eminently practical pursuit, even at this unpromising latitude, now that a row of splendid cast-iron and glass grape-houses had been erected in the famous Powerscourt demesne. Himalayan shrubs bloomed in this Gulf Stream—washed fold in the foothills of the Wicklow mountains. And now, in season, there were delicious Muscat and Trebbiano grapes for the Viscount's table. There were no pretensions to turn them into wine.

Lord Powerscourt had inherited the family passion for gardening. In the famine years of the 1840s his father, the Sixth Viscount, had laid out vast terraces to the front of the grey-stone Palladian house. The rock-moving labor at sixpence a day had provided relief for the distressed peasantry. Now twenty years later his son continued the grand project, travelling throughout Europe in search of baroque antiquities to adorn his Irish Xanadu.

It fell to Dunn to supervise the planting. The London nurserymen's catalogues bloomed with wonders. Botanist-explorers had scoured the Americas and the Orient for horticultural novelties – dispatching them homewards where necessary in clever sealed glass cases to be experimented upon and propagated.

For several years now rooted specimens of the fashionable icons of mid-Victorian gardening – wellingtonia and araucaria, rhododendrons and phormiums, exotic vines and climbers – had come trundling up the road in their crates to Powerscourt after a long journey by steamship and railway.

The work of landscaping and planting had gone on through that spring of 1867, even when an armed "Fenian" mob had stormed up the road from Dublin towards this Arcadian citadel of Anglo-Irish ascendancy. They had been routed by the constabulary. Two companies of Royal Marines remained billeted in the stables and at the Enniskerry crossroads should anyone try again.

This July morning it was not rebels who troubled Mr. Dunn. It was the condition of the viscount's vines. The bud-break of late spring had turned to rich green foliage. Now the leaves of one vine, then another, were rapidly turning red-brown, drying up and falling. "I saw clearly, that at the rate it was increasing, it would certainly soon destroy all the vines if it was not effectually checked," the gardener recorded. The two "late" grapehouses were rapidly infected.

Dunn noticed ugly blisters on the undersides of the afflicted leaves. Crumbled in the hand they divulged pale yellow blobs of evidently insect matter. He tried "the usual methods of exterminating plant insects . . . fumigating with tobacco and capiscums, dusting with snuff, sulphur, hellebore, Cayenne pepper &c.," which he found "of little use." He cut off and burned dying foliage. The disease progressed.

There had been mention that spring and summer in the horticultural press of "a new vine disease in the south of France . . . and in some of the London nurseries." Might this be the same affliction? Dunn did what any intelligent gardener of the period would have done (there was no state biological inspectorate), sending samples that August of his corrupted vine to the editor of the *Gardener's Chronicle and Agricultural Gazette* in London asking what might be the cause. He received a terse reply:

"Thanks for the vine-leaves. The disease is undescribed. We are preparing an article on it." The letter was signed simply "W."

It would take another fifteen months of anxious deliberations before the promised article by "W" (whom Dunn assumed correctly to be "the distinguished entomologist Professor J.O. Westwood of Oxford University") appeared in print.

3

The Ashmolean Museum, Oxford, October 1867

Of all the titles borne by mid-Victorian men of science, that of "insect referee" was one of the most curious. Schoolboys of the period might have taken unkind delight in setting beetle against beetle in flowerpot arenas to test the scandalous new biological theories of which their parents whispered blushingly. The position held by Professor John Obadiah Westwood, however, had nothing to do with arbitrating such enactments of the struggle for survival within species – the origin of which Charles Darwin had eight years before declared not to be the exclusive work of the divine Creator.

The professor, no friend of Darwin's theories, was both Oxford scholar and popular journalist – hugely popular in fact, as J. O. Westwood was a columnist for the *Gardener's Chronicle and Agricultural Gazette*. The journal's circulation had blossomed on the great mid-Victorian love affair with plants. Like the best editorial ideas, the premise of the professor's column was simplicity itself. If a member of Britain's eager army of amateur gardeners found something unusual scuttling in the conservatory, they might send it in a matchbox care of the journal to the professor – for the insect referee to pronounce in print on just what it might be. It was science for everyman, a voyage of the *Beagle* in your own backyard. The more unusual the discovery the better. By the mid-1860s, some most unusual things were turning up in the United Kingdom's graperies and melon-houses.

Professor "W" had answered his Wicklow correspondent abruptly. The following month, September 1867, he received a postal packet forwarded from a correspondent in Cheshire. It contained "leaves from a young vine planted the February preceding." The undersides were covered in a mass of blisters, the upper faces were minutely serrated with tiny bristle-fringed slits. When the galls were opened each contained a mass of minute eggs – and in some cases tiny yellow-orange insects. The Cheshire gardener was puzzled. What were they? Only one vine out of twenty-five growing in a large greenhouse had apparently been affected.

The professor had been here before. In June 1863 he had been sent by a grower in the west London suburb of Hammersmith "a vine leaf covered with minute gall-like excrescences, 'each containing,' in the words of my correspondent, 'a multitude of eggs, and some perfect Acari [mites], which seem to spring from them and sometimes a curiously corrugated Coccus.'"

He examined them under his dissecting microscope, and concluded that the galls were caused by "the irritation from the sucking of the leaf by the full-grown insect enclosed within . . . as in a living tomb."

The curious insect he thought belonged to the family *Aphidae* and not the *Coccidae*. But for "want of a knowledge of the male" the professor could comment no further. The corrupted leaf and scraps of mummified invertebrate had been filed away in a cupboard of the Ashmolean Museum. News of the west London discovery had gone unpublished.

Now four years later the unknown aphid had turned up again on vine leaves in Wicklow and the north-west of England. There was something else. At Powerscourt, Dunn had meanwhile thought to "turn up part of the border to examine the condition of the roots." He found them "attacked by the insect

in myriads, burrowing into the young rootlets . . . and destroying them." The professor was about to see the insect's underground work for himself.

A month after sending the blistered leaves to the *Gardener's
Chronicle and Agricultural Gazette*, the Cheshire correspondent sent a second parcel – this time containing rootlets from
the same vine that had been afflicted by the mysterious galls.
Professor Westwood examined them microscopically. They
were infested with what was clearly the identical insect to that
on the leaves. "In the latter [underground] mode of attack, the
perfect [adult] insect makes a wound in the delicate rootlet,"
he observed, "by inserting its rostrum into the wood . . . decay
is thus induced, which penetrates in the form of little cankerous
spots, and sometimes extends to the centre, cutting off the
supply of nourishment."

Whatever this tiny unknown aphid that infested both the
leaves and roots of vines might be, as a good entomologist, the
professor must call it something. In the spring of 1868, in his
words: "I proposed to the Ashmolean Society of Oxford to
name the insect *Peritymbia vitisana* in allusion to the tomb-
like gall on the leaves formed by the female insect."

Where had it come from? Professor Westwood was as perplexed as anyone.

4

Down House, Downe, Kent, 1868

Charles Darwin thought himself a bit of a dud at botany. He hardly knew "a daisy from a dandelion," he would tell visitors as he pottered in his Kentish garden. It was a refuge from the intellectual storm that still raged round the world nine years after the first publication of *On the Origin of Species by Means of Natural Selection*. It still blew a hundred penny-post envelopes a day through his letterbox and gaggles of curious pilgrims up the lane to the doorway of Down House.

In mid-life the bold explorer-naturalist of the *Beagle* had become a snuff-dosed invalid. His garden however became a laboratory as well as a retreat. Living specimens arrived in sealed glazed cases to be observed and experimented upon. The plant world came to him, borne by steamship from faraway colonial out-stations on the great mid-Victorian tide of horticultural obsession. As the creationist counterblast rumbled, Darwin toiled with his cutting trays and flowerpots, looking for ever more proof with which to confound his critics and perhaps subdue his own lingering glimmers of doubt.

"Plants are splendid for making one believe in natural selection as will and consciousness are excluded," he told the biologist Thomas Huxley, liveliest of Darwinism's champions. Plants, both commonplace and exotic, became the means to demonstrate the adaptive ingenuity of nature by the process of natural selection. He was surrounded by them.

A grape-vine grew against the side of Down House. Why did its flower-stalks show sensitivity to touch and residual power of motion? Why did the shoot of "a young Muscat grape under a bell-jar in the hot-house [make] each day three or four very small oval revolutions"? With its separate, function-specific tendrils and flower-stalks, in Darwin's view clearly derived from the same ancestral structure, the vine presented the evolutionist with "as striking an instance of transition as can well be conceived."

Further proof of adaptation was under his nose, especially in the wild orchids that bloomed in the chalky banks and shaded copses around Downe village. He became obsessed with *Orchidaceae*, in their multiple variations of form, color and complex sexual anatomy. Where fashionable society might see them as evidence of the Creator's bounty, to Darwin the gorgeous show was simply to better entice cross-pollinating bees. According to his definitive biographer, Darwin's "intention was to show that even the most complex structures and life-cycles, even those that depended on completely different organisms such as insects for their fulfilment, could be explained by natural selection."

Darwin's transatlantic champion, the Harvard botanist Asa Gray, found religious solace in considering the relations between insects and plants. In Gray's Christian outlook, such beneficial co-adaptations as insect pollination of flowering plants surely reflected the intent of the Creator. In their voluminous correspondence Gray argued with his friend that God introduced "favourable variants" into what the botanist conceded was the great scheme of evolution. Natural selection, naturally, chose them.

A less benign model also held true. Whereas bees and orchids lived in buzzing harmony, the plant kingdom bristled with spikes, stings, traps, barriers, disguises and poisons to repel predators and pathogens – viruses and viroids, bacteria, fungi, nematodes and insects. While the concept of natural selection

referred to variations within species which made one reproductive individual "fitter" than another, without the "stress" of
climate, environment and predators, there was nothing to test
that fitness. It worked both ways. Natural selection favored
those invaders better adapted to penetrate the defenses. Fungi
and insects were also engaged in the "struggle for life, entailing
divergence of character and the extinction of less-improved
forms," as the last paragraph of *Origin* expressed it. Darwin
called it the "war of nature."

In 1868 Darwin published *The Variation of Animals and
Plants under Domestication* in an attempt to resolve a fundamental question left unanswered in *Origin*: how did individual
differences in organisms come about? Without a compre-
hension of genetics, his observations centering on anecdotal
evidence from livestock breeders and correspondents of the
Gardener's Chronicle led to only partially successful conclusions. Looking for proofs, Darwin turned again to the grape-
vine, Vitis vinifera, tended by man as he said "since remotest
antiquity," selected from the wild and propagated by cuttings
into a myriad cultivated varieties.*

Darwin began at the beginning. "The best authorities consider all our grapes as the descendants of one species which
now grows wild in western Asia," he wrote, "which grew wild
during the bronze age in Italy, and which has recently been
found fossil in a tufaceous deposit in the south of France."

"The cultivated varieties are extremely numerous," he noted.
"Count Odart [author of *Ampélographie universelle*, first published in 1845] admitted the existence of up to a thousand. In
the catalogue of fruit cultivated in the Horticultural Gardens
of London, published in 1842, ninety-nine varieties are enum-
erated. Wherever the grape is grown many varieties occur: Pal-

---

* "Cultivars" of the vine in the botanical sense are "clones" arising from a single
seedling and then propagated by cuttings, so that all the original "mother-vine's"
descendants are genetically identical.

las describes twenty-four in the Crimea, and Burnes mentions ten in Cabool. The classification of the varieties has much perplexed writers."

Indeed it had. An ancient branch of inquiry, once the preserve of monastic savants, had become newly fashionable in the first half of the nineteenth century. It was called "ampelography" (*ampelos*: vine), a Greek derivation for what had become an almost exclusively French pursuit – to bring botanical order to a chaotic, squabbling, overbred, extended family: the cultivated varieties of the vine.

Like Charles Darwin, these distinguished gentlemen on the other side of the English Channel worked from the proposition that "all our grapes" sprang from a specific near-Oriental ancestor. All the grapes that mattered, that is. What occupied the "ampelographers" was classifying the cultivated varieties that millennia of human intervention had teased from this single species – Vitis vinifera. They were very important. They made wine.

5

❧

European civilization had been transformed by the culture of the wine grape, the cultivated descendants of Vitis vinifera, which had spread westwards from central Asia borne on Phoenician galleys and Roman ox-carts. By the time Darwin observed the revolutions of his little Muscat grape-shoot, European society had been again transformed over three centuries by botanic gifts from the Americas – maize, tobacco, the potato. That the New World was full of vines had long been known. Eleventh-century Scandinavian voyagers nosing westwards in longships from Greenland had found wild vines growing abundantly on a strange new shore. They bore *vinber* – grapes, familiar at least to those adventurous Norsemen who had encountered the vines and wine of Iberia and the Black Sea. They called their short-lived new home "Vinland."

Post-Columbian explorers had also found wild vines. Unlike tobacco, these had not divulged a golden narcotic secret when transplanted home. Exotic species from the colonies were brought by sailing ship as rooted cuttings or seeds (wild vines from New England, Vitis labrusca and Vitis vulpina in their post-Linnaean classification, are recorded in a London herbarium of the 1650s), but remained mere botanic curiosities. They proved decorative enough in gardens and bore coarse fruit. Any fermented outcome was repugnant.

In America itself however several of these native vines, attuned by evolution to withstand humid, mildew-bringing summers and freezing winters, had been domesticated into wine-yielding varieties. Serious efforts had begun in the late eighteenth century. The results were passable enough. In Ohio a thriving little industry had been made out of Major John Adlum's Catawba. Then there was the "Schuylkill" grape, reportedly discovered near Philadelphia and named after a river in Pennsylvania by a certain Mr. Alexander, William Penn's gardener. The taste of its fermented juice was likened by some to that of a "red Bordeaux."

The benefits of Herbemont (also called "Warren," having been found growing wild in Warren County, Georgia) were first proclaimed in South Carolina in 1798. In 1818 "Isabella" was found as a seed in a garden at Dorchester, South Carolina, by a Mrs. Isabella Gibbs of Brooklyn. It was cultivated by the noted nurseryman W.R. Prince who named it in her honor. The "Cunningham" grape was first made into wine by a Dr. Norton in 1835. "Concord" originated from the seeds of a wild grape planted in 1846 by Ephraim W. Bull at Concord, Massachusetts. The origins of the variety known as "Jacquez," also known as "Black Spanish" or "Cigar Box," are obscure, but cultivated in the southern states of America it made a passable wine, dark in color and high in alcohol. Later ampelographers consider Jacquez to be an accidental hybrid of two American species and imported vinifera.

All this effort was as much necessity as frontier spirit. There was nothing else to drink. Old World vinifera varieties had been brought westwards by waves of colonists since soon after the first landfalls. The earliest French settlers had tried to grow grapes in Florida and Illinois, Huguenot refugees had tried in New Paltz, the English (they hired Frenchmen to do it for them) had planted vines in Virginia. Apples, pears and peaches flourished in these hopeful simulacra of home. European vines withered and died.* Only the brave or the foolish had kept on trying.

In 1802 the French botanist André Michaux visited "First Vineyard," a utopian project established twenty-five miles north of Lexington in Kentucky by John-James Dufour, a Swiss who had come to America in 1796 with a vision of founding a wine industry in the New World. He spent two years looking for the right site before summoning his family and adventurous neighbors to follow him from Vevay, north of Lake Geneva.

He found his promised land on the Great Bend of the Kentucky River, where the topography and climate seemed ideal for the cultivation of the vine. Michaux noticed "delicious summer grapes" growing wild in abundance. He judged the calcareous soil similar to "the best vine growing regions of France." Dufour and his followers had devoted "unremitting labour," Michaux noted, to making a collection of twenty-five varieties – both European and native vines – which were "planted fixed with props similar to those used in the environs of Paris." When Michaux returned two years later, the harvest had failed and failed again. With the onset of summer heat, the vines were engulfed by mildew which had rotted the bunches on "all but three or four species." The survivors were devoured by huge flocks of birds.

Dufour refused to give up. He struggled on with Catawba vines as his despairing followers trickled away to Indiana to try again, this time with native Schuylkill grapes at "Second Vineyard." The wine proved disastrously unsaleable. Dufour nevertheless eventually joined them. In 1825 he published a

---

* According to Major John Le Conte of Philadelphia, writing in the 1857 Report of the U.S. Commissioner of Patents: "About sixty years ago, there was scarcely a yard in the city of New York which did not possess foreign vines producing fruit of the finest quality. In the garden belonging to the house in which Colonel Aaron Burr lived, about the year 1793, at the corner of Nassau and Cedar streets, there was the finest . . . collection of grapes I ever saw. All the choicest varieties that would be found in Europe flourished there . . . Now there are none they will not grow there."

famous work on the culture of the vine, *The American Vine Dresser's Guide.*

The number and variety of native species was astonishing. The Turkish-born naturalist and pioneer American ampelographer Constantine Rafinesque wrote in 1830:

> *The subject is new and obscure. The botanical species are scarcely identified and their numberless varieties have been overlooked. I have ascertained about forty species and a hundred varieties but I must confess it is not easy to say whether one or the other . . . many species have no doubt escaped my researches, they abound in the woods since the seeds do not always produce the identical kind, and Major Adlum has stated to me to have seen two hundred varieties at least.*

As the frontier pushed outwards into the vast south-west there were more discoveries of unreported species and natural hybrids. In 1826 the young Frenchman Jean-Louis Berlandier was dispatched by the Philosophical Society of Geneva on a botanical mission to Mexico. He made drawings and collected thousands of plants – but fell out with his Swiss mentors when his vaulting ambitions outran his brief. He refused to return when ordered to, settling down with a Mexican woman and studying native vines and cacti, the fermented juice of which were both used in native religious ceremonies. Several mystic beverages are described in his journals.

Attempts by European colonists to produce their own comforting liquors from vinifera varieties had "not prospered and the wine was of mediocre quality," Berlandier noted. The wines of Potosi had "too much alcohol, are too pale and go easily to the head," he wrote in 1834. But "without the introduction of an exotic plant [the European vine], there exist in the inhabited regions of Texas three grape-bearing species, two of which undoubtedly belong to the genus Vitis and which furnish quite good grapes every year."

He saw "one species with white fruit growing along the banks of the Arroyo de León and whose taste is quite agreeable. The species with red fruit is much more abundant and I have drunk wine from its grapes which retained a sweet-sour flavour even in maturity. Perhaps cultivation could improve these two species . . ."

The botanist despaired meanwhile that "the inhabitants of Béxar (even some Anglo-Americans) rarely talk about a well-cultivated field or a splendid harvest, rather it is said here or there – in alluvial terrain covered with fossils – there is a gold or silver mine."*

It was no accident that so much of this plant-hunting energy was francophone. The nation bred botanists as Prussia bred chemists. Botany was a key science in the National Museum of Natural History in Paris, and regional institutions such as the Botanic Gardens of Montpellier in the south were world class. Agriculture was the primary occupation of seven Frenchmen in ten, and at its heart was an activity which delivered four times the value of any other crop. It could, as Professor Guyot would note in his Exposition report, extract high returns from the poorest soils. A few pebbly hectares could support a peasant *vigneron* and his family with a cash crop. In this bounty lay a social vulnerability.

By the 1840s the vine and its products delivered one-sixth of France's state revenues, wine was the nation's second export after textiles, and a third of the population derived a living from it. The culture of the grape and the making of wine was the very soul of the nation. If anything stirred in the world of the vine, there had to be French fingerprints on it.

The end of the Napoleonic wars allowed American vines,

---

* Berlandier died in the summer of 1851 while trying to cross the San Fernando River. His notes and specimens ended up in the safekeeping of Asa Gray in Washington.

wild and cultivated, to be freely shipped across the Atlantic. The celebrated collection of the Jardin du Luxembourg in Paris contained twenty-three varieties by 1825. The catalogue of the Audibert Frères, nurserymen of Tonelle near Tarascon in the Rhône delta, offered twenty for commercial sale in 1831. Keen *viticulteurs* began to experiment.

B.-A. Lenoir, author of *Un Traité de la culture de la vigne et de la vinification*, suggested in 1828: "There are some wild vines especially in America, which give rather good fruits, which cultivation would make even better. These species might perhaps communicate by 'grafting' their robust temperament to our vines."

Some were keen to find out. Louis Cazalis-Allut, president of the Central Agricultural Society of the Hérault, began growing Isabella in his *domaine* at Aresquiès near Frontignan in 1832. But, he noted, it made a wine with such an overwhelming taste of *framboise* as to be quite "nauseating." It might be blended, however, with local product and, he discovered, be "grafted" successfully onto native European roots – producing grapes that were both "delicate and pleasant to eat."

The Isabella vine proved especially popular in Europe. The Marchese Ridolfi imported large quantities into Tuscany in the late 1840s as a horticultural curiosity, and would do so again a decade later for a different reason. It grew rapidly into luxuriant decorative bowers to adorn fashionable gardens. The exotic vine bore alluringly plump grapes but, as Cazalis-Allut had discovered, its wine-making potential was unpromising. Count Odart came straight to the point in the fourth edition of his ampelography (1859), "I shall be reduced so to speak about the very small number of vines of America which might have some merit." He had corresponded, he wrote, with "several French proprietors . . . who think like me that the Isabella grape with its medicinal taste of cassis is good for nothing."

A certain M.M. Tourrès, *pépiniériste* of Touraine, had lately been extolling his imported samples of "Scuppernong" grape (the Vitis cordifolia discovered by André Michaux) as "better than anything we possess in Europe." Count Odart dismissed the insipid Scuppernong as "completely without flavour." He conceded however that wine made from American "Katawba" grapes might prove more acceptable.

A Bordeaux vine-grower, Léo Laliman, was especially enamored of the promise of America. From 1840 onwards this flamboyant former cavalry officer in the army of King Louis-Philippe received samples of Scuppernong, Catawba and other varieties shipped across the Atlantic from nurseries in Georgia and Pennsylvania, altruistically dispersing them to fellow enthusiasts around the Médoc and Saint-Émilion.

Little by little the American visitors spread happily across the wine regions of France – ampelographical conversation-pieces tucked in vineyard corners and nurserymen's plots. On they went to collectors in Italy, Germany, Austria and Iberia. They were biologically fascinating. They could be artificially "hybridized" by growing new plants from seed after cross-pollinating and, as Louis Cazalis-Allut had discovered, different species of Vitis could successfully be grafted root to "scion" to make not a hybrid but a compound plant that seemed to function perfectly well. Many more American vines were on their way to Europe.

In the late 1860s grape-vines in south-east France began to wither and die for no obvious cause. Not until an investigative committee dug up an apparently healthy vine and found its roots encrusted with tiny yellow aphids was the villain revealed. There were many who thought differently – that the "phylloxera" was not the cause but an effect of some other mysterious malady – and the scientific detective work would take many more years. This illustration dates from 1874.

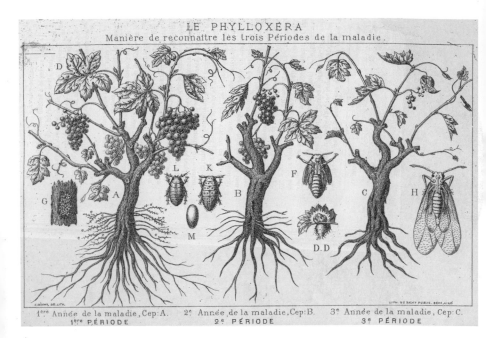

LE PHYLLOXÉRA
Manière de reconnaître les trois Périodes de la maladie.

1ᵉʳᵉ Année de la maladie, Cep: A.   2ᵉ Année de la maladie, Cep: B.   3ᵉ Année de la maladie, Cep: C.
1ᵉʳᵉ PÉRIODE   2ᵉ PÉRIODE   3ᵉ PÉRIODE

By the mid-1870s the phylloxera's life-cycle was more or less understood, and France's vine-growers had been mobilized in "study and vigilance committees" armed with recognition charts of the invader.

Attempts to eradicate the plague by uprooting and burning infected vines were too little and too late. It was no easy task to extract a deep-rooted vine from the ground – windlasses and pulleys were used, or an outsize crowbar called "the goat." Later this wheeled crane was devised.

The phylloxera story had its heroes, who had to fight not just the aphid but indifference and ignorance among vine-growers and the political establishment. The Montpellier botanist Jules-Émile Planchon (*above, left*) was one of the first in the field, realizing very quickly that the aphid was the cause of the disaster – and, crucially, that it had somehow come from America. Charles Valentine Riley (*above*), the English-born state entomologist of Missouri, had simultaneously found the insect on the roots of vines in the United States, and came to France in 1871 to prove the connection. He and Planchon were among the first to suggest that American vine species were "resistant" to the insect (although they made some serious errors as to exactly which ones). Professor Georges Balbiani (*left*) of the Ècole de France investigated the curious sex-life of the insect. His resulting winter-egg theory led to another diversion in finding effective means to combat the plague.

Ground zero: 21, rue Longue in the little town of Roquemaure in the Gard. In 1863 a wine merchant named Borty planted some rooted vines sent to him in a sealed box by a friend in New York. The following year vines in the fields around began to wither and die mysteriously. Planchon later traced the first outbreak to the walled vineyard behind Borty's modest establishment.

6

By the third decade of the nineteenth century the plant-hunter as hero-explorer was being supplanted by the botanist as entrepreneur. The planet seemed a vast vegetative *galère* which could be plundered for fashionable novelty and commercial gain. Colonial economies might be transformed by the introduction of a strategic crop, if specimens could be plucked from their place of origin and transported successfully. In 1831, for example, a young Englishman named James Busby advanced a plan to cultivate vines on a commercial scale in New South Wales. Where better to take advice on what varieties might prosper on the wild colonial shore than from the director of the Botanic Gardens of Montpellier in the department of the Hérault in south-east France, renowned center of ampelographical inquiry.

"Professor Delisle received me with great kindness," Busby recorded in his journal, "and told me I was welcome to cuttings of all the vines he had. I received from him an introductory note to M. Audibert, the proprietor of a very extensive nursery at Tarascon [whom] he recommended me to consult particularly on the best mode of preserving the cuttings from frost and damp."

Urban Audibert indeed had very practical ideas on how to transport vines across the world. "Boxes were lined with double-oiled paper, to prevent the access of air and the escape of humidity," recorded Busby. "Moss, having been slightly watered was stuffed in at the ends of both bundles of plants . . . and the cases closed.* This is the mode adopted by Messrs. Audibert Frères in sending vines to Russia." And, it might be assumed, how the industrious brothers in turn received exotic vines from America.

The great botanic chess game was given a boost with the invention in 1835 of the "Wardian Case," the glazed terrarium devised by Dr. Nathaniel Ward, physician, of Wellclose Square, London. Hermetically sealed, with enough carbon dioxide production to maintain growth, it protected delicate living specimens as they migrated across oceans. The botanist Joseph Hooker was among the first plant-explorers to use the new cases, shipping many new specimens back to Britain during his four-year voyage to the Antarctic acting as assistant surgeon with Captain James Clark Ross, from 1839 to 1843. Robert Fortune used the cases to transport twenty thousand tea plants from Shanghai to Assam in north-east India. Charles Darwin had one in his drawing room. The mid-Victorian orchid craze that stripped the coastal jungles of Brazil was made possible entirely because of Dr. Ward's case. There were no barriers, no inspectorate, no concept of biological quarantine. Then things began to go wrong.

In the summer of 1843 the potato crop in a five-state area around Philadelphia and New York corrupted into rotting mounds. It happened in a matter of days, and although such diseases had come and gone before, this time foliage, under-

---

* The vine cuttings from Montpellier and Tarascon were first sent to the Royal Botanic Gardens at Kew, where cuttings of cuttings were taken. They were repacked in their oil-paper-lined boxes and eventually reached Sydney "in the highest state of health and vigour" on a British convict ship.

ground tubers and harvested potatoes stored in clamps and cellars were all stricken. Two summers later, potato beds in Belgium, Germany and northern France turned black. John Lindley, professor of botany at University College, London, ascribed it to the heavy rains of the late summer – the potato had become charged with water and suffered a "dropsy," to rot thereafter. The professor was not stupid. Without an understanding of genetics, or Louis Pasteur's as yet unarticulated "germ theory of disease," plant pathology was an observational science. If it rained to excess, that was the cause.

That summer of 1845 the new "potato murrain" was observed on a Northamptonshire smallholding by an English curate, the Reverend Miles Joseph Berkeley, a keen mycologist who had published pioneering work on the causes of plant disease. He noticed a minute whitish fringe around the decaying leaves. In the face of establishment scoffing, the Rev. Berkeley proposed that a fungal organism was the cause of the rot, rather than its consequence. In September the blight reached Ireland.

Just where *Phytophthora infestans* (technically a fungal-like oomycete) first embarked on its journey of destruction still remains uncertain. An amateur botanist might have brought infected plant materials from Mexico (another theory favors the northern Andes) into the United States around 1840, and the blight have subsequently been transmitted to Europe on an infected tuber.

Modern science has tried to put together what happened. For centuries vegetative propagation, the use of cuttings rather than seed to reproduce an individual plant, was the favored means of potato cultivation – just as it was with the vine. By the 1840s, therefore, the potato crops of Europe consisted of vast clones of genetically uniform material, spatially removed not only from the center of origin of the plant itself, but also from its most dangerous enemy. Bringing them together would bring disaster.*

Another North American visitor was about to make its landfall. In 1834 cultivated vines in the eastern United States were affected by a greyish, dusty fungal growth that began its attack on the shaded underside of foliage. Leaves were observed to turn upwards as if in supplication. Soon the whole plant was affected by "powdery mildew." The crop was ruined, the American plants withered, but did not die.

In the summer of 1845 an English gardener, Edward Tucker of Margate, Kent, discovered the leaves of a grape-house vine smothered in mildew. It rapidly spread to the outdoor trellises adorning his employer's seaside villa, covering them in what one account called "a dusting of white and pulverulent meal; it spread rapidly onto the grapes themselves, withering the bunches when they were small and green, or causing the grapes to crack and expose their seeds . . . The disease was accompanied by an unpleasant mouldy smell, and it ended in the total decay of the fruit." The Rev. Berkeley interrupted his work on the potato blight to investigate. It was a new parasitic organism, he deduced, a fungus previously unknown in Europe. He did not comment on where it might have come from. He called it *Oïdium tuckeri*.†

The fungus almost certainly reached Margate the same way *Phytophthora infestans* arrived in Northamptonshire, carried from America on a botanical sample. The Isabella vine, which had been imported into and disseminated all over Europe as an ornamental curiosity since the 1830s, is the most likely suspect. The following year the hateful *oïdium* turned up in the

---

* Retrospective analysis has led to the hypothesis that the Irish potato famine was caused by a pathogen population of extremely limited genetic diversity. Thus a chance genetic "migration" of even one or a few clones of the late blight pathogen had devastating consequences.

† It is now cited as *Uncinula necator*.

graperies of King Louis-Philippe's royal kitchen-garden at Versailles and a little later in Baron de Rothschild's grape-house at Suresnes, outside Paris. In 1851 vines in the department of the Hérault were afflicted.

The infestation would begin on one or two vines, typically in the shade of a tree or building (the fungus liked low light conditions), then would spread rapidly through the whole plantation. Three years later the *oïdium* was in every vineyard in France. Many were simply abandoned. A number of venerable vineyards around Paris were turned over to growing vegetables for the markets of Les Halles. A prize of twenty thousand francs was offered by the French Society for the Encouragement of National Industry for a cure. Crackpot remedies abounded – herbal bonfire smokes, douching the roots in brine, washing foliage with soapy water, distempering with bizarre chemical cocktails, planting potatoes among the vines to somehow draw away the poison. None worked.

As with the potato blight, debate raged on whether the mildew was the cause of the disaster or the symptom of some other depredation. And while some varieties were blighted immediately, others seemed less vulnerable. Eventually, though, all succumbed.

French wine production collapsed from an annual thirty-nine million hectoliters to eleven million. The price of wine soared. The "Midi," the Mediterranean south, was particularly afflicted as wind-dispersed spores blew from vine to vine in the dry heat in which the organism thrived. By the summer of 1854 the whole of southern Europe seemed to be facing a biological catastrophe.

"The little vineyard that for centuries yielded its petty revenue, which in storm and sunshine bravely struggled to support the family, is exhausted," reported *Cozzens' Wine Press*, a lively newsletter published by the New York liquor merchant Frederick S. Cozzens. "Nature can do no more, and in sorrow and poverty, the peasant must turn to face other employments. So far the northern countries have been less affected, but in the south of France, Spain and Portugal, in Madeira and Italy, prospects seem to favour the conclusion that in the course of a few years, these famous wines will be at an end," Cozzens predicted apocalyptically.

From Oporto came ominous news: "The Douro country is utterly ruined," wrote the *Wine Press*'s Portuguese correspondent, "and I much fear that the bad smell which the vines exhale may generate some devastating epidemic." The correspondent of the *Newark Daily Advertiser* reported from Genoa: "The vine again shows symptoms of the disease that has cut off the grape-harvest, which is to Italy what the potato crop is to Ireland. It is not surprising that the ignorant peasantry should regard it as a divinely-appointed scourge, beyond the reach of natural remedy, but it is strange that it should have excited a prejudice against railroads."

Just so. The laying of "long lines of iron through the soil" was apparently being widely blamed for the disaster, while other enraged farmers ascribed "the vegetable malady to the smoke and gas of the locomotives," or to evil emanations from the telegraphic wires that accompanied the new-built track. In Tuscany "several miles of railway had been torn up and dumped in ditches." The alarmed authorities in Florence urged priests to provide an alternative explanation – the wrath of God for mankind's sins would suffice.

The New Jersey reporter's own analysis of the disaster's cause was more enlightened. "The grape is not indigenous to Europe," he wrote. "It is of eastern origin, and if historians be correct, a native of Persia. It is propagated by cuttings . . . year after year, century after century . . . This mode of cultivation, contrary to the method of nature, in all probability has exhausted the powers of the vine to reproduce itself."

This "degeneration" theory would be voiced again when the phylloxera began its mysterious underground predations.

"It is remarkable also that a similar disease has affected the potato, a plant propagated in the same way, namely, by cuttings instead of seedlings," the clever correspondent continued. "The cause of the disease, therefore, may be ascribed to the method of cultivation, and the only remedy is to begin again by planting from the original vine in its wild state, or transplanting to Europe the vines of America."

The journalist did not know it – neither yet did anyone else – but the cause of the disease was all those Isabella vines imported to adorn Tuscan villas. They had carried the fungus but themselves been resistant.

Redemption would not come from America. Not this time. It came from a group of French botanists, zoologists, chemists and *viticulteurs* who were determined to stand and fight. Science would deliver an answer, just as it had two decades before when a voracious caterpillar had threatened the vineyards.* The attacker this time was clearly some sort of fungus. It had been known by horticulturists since the 1830s that application of sulphur was a remedy for such parasites, though no one quite knew why. Greenhouse experiments using the chemical sprayed in a water solution proved promising, but how to do it in the field? Dusting with fine powdered sulphur manufactured on an industrial scale and transported by rail was the answer.

For five years the vineyards of France bloomed with dusty

* The "Pyrale," the leaf-munching caterpillar of a nocturnal moth, had disastrously appeared in the vineyards of southern Burgundy and the Champagne in the mid-1830s. The government commissioned Victor Audouin, professor of the Museum of Natural History of Paris, to study the parasite, while many vinegrowers sought their own more or less scientific remedies. The Beaujolais *vigneron* Benoît Raclet famously discovered a simple cure by scalding vines with boiling water.

explosions as women and children spluttered under sacks of eye-watering chemical,* feeding toiling lines of laborers. Special bellows were devised, riddles, "dredge-boxes" and patented tin funnels tipped with braids of wool like giant sea anemones to get the sulphur onto the vines. One man working seven hours a day could apply ten kilograms of sulphur, noted a *Times* correspondent in the Midi – five days per man per hectare. It had to be done three times: when the first shoots appeared in spring, at flowering, and when the grapes were just formed. It was a colossal operation, breaking the biological siege – but it worked. By 1858 the *oïdium* was in general retreat, although it lingered in some spots such as the Médoc until 1863.

Not all *viticulteurs* were persuaded. Some complained that the spraying affected the taste of the wine – it actually made it better, said the chemical's promoters. The practical Henri Marès of Montpellier published a propagandizing pamphlet arguing that fruit set was improved after dusting, and ripening of the grapes came earlier – cutting the risk of harvest damage. The operation should be carried out no later than a month before the *vendange*, but there were inevitable residues. "The small quantity of sulphurated gas that is normally to be found in the new wine is considered valuable to preserve it," noted the Cincinnati oenophile William J. Flagg.

Sulphurization spread into the most obscure corners of Europe and North Africa. In that curious vine-growing Aegean outpost of British dominion, the Ionian Islands, "rough brimstone" was employed. The poorest Spanish peasants used dust from the roads. The chemical reek in the broiling midsummer vineyards of Algeria was said to be especially noxious.

---

* The Imperial Inspector General of Agriculture recommended the use of "goggles similar to those worn by railway locomotive drivers." *The Times* reported on the effects of sulphur dust on the tender eyes of women and children, "subject to ophthalmia, inflammation of the conjunctiva," inviting readers to contemplate the suffering entailed in maintaining their wine-drinking pleasures.

There was something else. Those enlightened vine-growers who had planted American species noticed that they did not succumb at all to the mildew's embrace. Not just in France. In 1857 the Accademia de Geogofila of Florence reported "the curious fact that the grafting of American vines on those of Tuscany produces a great increase in the quantity of grapes and that vines so grafted are immune to the *oïdium*." There was a drawback: "The wine though abundant is much inferior."

The phenomenon of American immunity was commented upon in the newsletters of French agricultural societies, many of them newly established to fight the common fungal enemy. Charles Darwin interpreted such reports from the front as exemplars of evolutionary adaptation. "During the vine disease in France certain old groups of varieties have suffered far more from mildew than others," he wrote in 1862. "Thus the group of Chasselas, so rich in varieties, did not afford a single fortunate exception; certain other groups suffered much less; the true old Burgundy, for instance, was comparatively free from disease, and the Carminat likewise resisted the attack. The American vines, which belong to a distinct species, entirely escaped the disease in France," he noted perceptively, "and we thus see that those European varieties which best resist the disease must have acquired in a slight degree the same constitutional peculiarities as the American species."

It was already suspected by some that the powdery mildew had come from the United States. American species growing happily in France and Italy escaped, while vinifera succumbed like Polynesians to the common cold. Why? Believers in evolutionary theory could put together an answer. It was about adaptation. It was about Darwin's war of nature.

7

France's intellectual establishment were engaged in a war of their own. The great scientific debate in the 1860s concerned whether life could be created in a soup of chemicals in a laboratory test tube. A doctor from the Jura named Louis Pasteur argued it could not – that every living being, even the smallest "germ," had to be produced by another. His startling "germ theory of disease" had sprung from observations of what caused fruit liquors to ferment in the first place. Wine, it seemed, was what micro-organisms did to grape-juice.

When Charles Darwin's theories first crossed the Channel they were met by a "conspiracy of silence."* The evolutionist's ideas were dismissed in 1864 by Professor P.-J.-M. Flourens, president of the Académie des Sciences, as being "deficient in the basic rules of logic"; polite society deferentially followed his lead. Catholic opinion found the whole thing repellent. Even radical republican followers of the ideas of Auguste Comte, apostle of positivism and science-based progress, preferred home-grown versions of *le transformisme* to explain the way the natural world was ordered.

French botanists, however, would prove readier to embrace Darwinism. As it was for their British counterparts, it was eas-

---

* The phrase was Thomas Huxley's. *The Origin* was first published in French in 1862 in an unauthorized translation by Mme. Clemence Royer, a Geneva-based political refugee from the Second Empire who added many curious ideas of her own.

ier perhaps for them to calmly contemplate the idea of natural selection in vines and orchids than among chimps and cavemen. French geologists too had been demonstrating with their rock-hammers since the 1800s that diluvian theology was not enough. Those strange fossil imprints of plants and animals had not been placed in the ground by the Divine Creator as an impish test of faith.

The world, it might now be admitted, was very old. So too it seemed was the grape-vine. There was great excitement in 1865 when the geologist E. Munier-Chalmas found a fossil vine, near Sézanne in Champagne, which flourished in subtropical forests during the Lower Eocene (fifty million years ago). It was borne in triumph to the Sorbonne. As imprints of ancient vine leaves and petrified seeds emerged from the rock, as the astonishing diversity of American (and new east Asian) species began to be understood, the ampelographer himself had to adapt to survive. A new kind of science took over. From the early 1860s onwards, taxonomists strove to delineate not just what man had wrought with the grape-vine, but what evolution had wrought in the first place. They have been arguing about it ever since.

After a century and a half of debate, the family *Vitaceae* to which the genus Vitis belongs* is now considered to have originated from an unknown ancestor in Asia around 140 million years ago, distinguished by its capacity to climb the trees and pierce the leaf canopy of Cretaceous forests. Differentiation of the genera – there are eighteen in the modern *Vitaceae* – is thought to have occurred in the Mesozoic era, before the separation of the continents. Four of the genera bear edible

---

* The French botanist J.-E. Planchon in his definition of 1887 split the genus Vitis into two "sections," the Muscadinia, with three species, indigenous to North America; and the Euvitis, containing nineteen species, including American and Asian vines and the "European" vine, or Vitis vinifera. Hybrids within sections are possible, but not across them. In 1967 the distinguished French ampelographer Pierre Galet listed sixty species of Vitis.

fruits – Amepoleosis, Cissus, Testramigma and Vitis – the grape-vine.

The break-up of the last land bridge in the Quaternary period split the Northern Hemisphere–spanning range of primitive Vitis into two – American and Eurasian. The further "speciation" of the oceanically divided genus was accelerated in the ice ages that followed as populations survived in pockets protected by topographical chance from the glaciers' cold footprints. In North America that process was highly differentiated. In Eurasia it was not. As Darwin understood, it was indeed a benign pocket of Transcaucasia that spawned the single species, Vitis vinifera, around 1.7 million years ago.

Wherever Vitis blinked back into the rising warmth of the Pleistocene epoch, on good evolutionary principles, the retreat of ice-sheets meant re-selection for adaptation to the new climatic and geological environment. It also meant adaptation to better survive pathogens and pests – viroids, viruses, bacteria, fungi, nematodes and insects, themselves embarked on their own evolutionary journey. It meant bio-diversity.

To French vine-growers of the 1850s, struggling to put their industry back together after the *oïdium*'s attack, all this would have been atheistical hocus-pocus. It was only half-understood why, but sulphur beat the fungus. Sulphur was expensive – up to one hundred francs for enough to treat a hectare – although the use of child labor kept the price down. American vine species, familiar now for three decades, did not need all the fuss and expense – but could they make wine? In 1860 the indefatigable Léo Laliman published a book recommending his beloved *vignes américaines* to "all lovers of progress."

"Attempts to cultivate European vines were fruitless," he wrote of the mournful history of vinifera in the New World, "but Yankee character is to persevere and native vines were cultivated with great success . . . Catawba was preferred because of its resistance to frost. Scuppernong and Isabella came later and have shone with not less brilliance! Not one of these species had yet suffered from the *oïdium*," he proclaimed. "American vines remain virgin under the malady's embrace . . . It remains to be seen whether taste and flavour can be made appealing to European palates," Laliman had to admit, "but if these efforts succeed they will be like hens laying golden eggs."

His concluding lines were remarkable: "By grafting our European stocks with American wood, or grafting American roots with the wood of our varieties, there is no doubt we shall be victorious over the *oïdium* . . ."

With the defeat of the fungus by sulphur, however, the merits or otherwise of American vines became an argument on the sidelines. From 1860 French viticulture entered a golden age. From Perpignan to Dijon, from Bordeaux to Lyon, the railway-building boom had connected all the wine-region capitals to Paris. The viticultural landscape changed. Where bad wine from unforgiving soil had been made for want of anything better, the vines were abandoned. Where peasants had drunk *piquette** or some locally distilled spirit, now on high days and feast days at least they drank commercially manufactured wine. Beer was something for the bourgeois.

Methods of cultivation improved, machinery was introduced, human labor began to be supplanted by animal. Profits went up and so, dramatically, did the hectarage planted, especially in the four departments of Languedoc and Roussillon. The Mediterranean Midi – now able to transport cheap wine by rail to the thirsty cities of the north – was in effect given

---

* *Piquette*: "second wine" made traditionally for wine-workers by letting water and sugar ferment over crushed grapes and skins after the juice has been extracted. Adulterators notoriously added the aniline dye "fuchsine," containing arsenic, to transform *piquette* into "wine."

over to a vast monoculture, producing a third of France's total. The annual consumption of wine per head of the French population was fifty liters in 1848. By 1881 it would be eighty liters. It was a happy time too for the grand proprietors. In the Médoc especially, the vineyards so exquisitely graded in 1855 by the Bordeaux Chamber of Commerce had become baubles for the new rich of the Second Empire. Once parish names were commonplace for a generic wine, and labelling was a matter for the shippers. Now every owner, it seemed, had to have a "château" and a label announcing its product's place in the order of precedence. A convenient glue that fixed paper to glass had recently been invented.

A rash of miniature white-stone palaces – turreted in *faux*– Louis XIII style, downsized Louvres and Palladian villas – sprouted among the fields of quartzy stone where bourgeois families would descend each autumn to enjoy the bucolic delights of the harvest. Behind the mock *ancien régime* façades, vat-houses were equipped with modern presses, tanning drums and fermenting vats. The all-important *chais* were extended as prestige-conscious owners brought the business of ageing wines in barrels, long the preserve of Bordeaux merchants, within their own walls.

With the best vineyards offering a return on capital of up to 10 percent, there was a speculative boom. Château Lafite was bought in 1868 by the financier Baron James de Rothschild for four million francs. A decade later Château Margaux was bought by Comte Pillet-Will, governor of the Bank of France, for five million francs (in 2003 it was put on the market for €400 million).

In the unpretentious Midi it was different. "Châteaux" were hardly heard of; a vineyard was a *mas* – the old Provençal word for a farm. The grander owners were doctors, vets, lawyers, magistrates, tax-collectors, engineers and politicians, open to the atheistical precepts of science. Their estates on the sun-baked plains were worked by tenants, sharecroppers and, at harvest time, by the itinerant laborers known as *montagnols*. In the scrubby uplands meanwhile peasant proprietors tended a mosaic of tiny smallholdings, getting their produce to market, in one description, "borne in indian-files of carts painfully drawn by the little oxen of the mountains."

In 1868 the redoubtable Dr. Jules Guyot completed his monumental two-thousand-page survey of the French wine industry, *Étude sur les vignobles de France*, an oenological Domesday Book based on observations made over almost a decade. "Wine is the most precious and the most energy-imparting part of the diet," wrote the doctor. "Its use in family meals saves a third of bread and meat, but more than that, wine stimulates and strengthens the body, warms the heart, develops the spirit of sociability; encourages activity, decisiveness, courage and satisfaction in one's work."

His vision was global. "Today it is Europe that knows wine best but America and Oceania already understand the advantages of viticulture; and soon perhaps, having laid aside the sad prescriptions of Islamism, the peoples of Asia and Africa will turn their vast wildernesses into rich *vignobles* and thus rescue themselves from their indolence and fatalism."

The doctor could not praise wine enough. When it came to its economic importance, his report on viticulture as reflected at the Exposition Universelle was more practical. It provided this useful snapshot:

> *The vineyard occupies in France about 2,500,000 hectares, almost half the world's total, representing the sixteenth part of our cultivated soil.*
>
> *The culture of the vine maintains six million farmers and about a further two million suppliers, manufacturers,*

carriers and storekeepers. The vine is cultivated in seventy-nine departments. In some, such as the Gironde, the Hérault, the Charente, it occupies from 100,000 to 150,000 hectares. In most of the departments the products of the vine are much more valuable than anything else grown, increasing everywhere to an extraordinary degree the incomes of proprietors both big and small.

The golden age was about to be eclipsed.

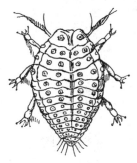

8

It began in the small town of Roquemaure in the department of the Gard on the right bank of the river Rhône, where an obscure wine merchant named M. Borty tended a smallholding of vines in a walled garden behind his modest establishment. In 1861 a "friend from America," a fellow vine-grower called M. Carle, had visited M. Borty. He promised to send back from New York some of his own native vines – "not for the quality of their wine," but so that his host "could discover for himself what their grapes were like." It was simple amateur botanic curiosity. M. Carle kept his word. "One day in the spring of 1862 a case arrived unexpectedly from America without any notice, the realization of a promise which M. Borty had never taken seriously and long ago forgotten," according to a little exhibition in Roquemaure's tourism office. The case contained rooted vines, carefully labelled by his émigré friend with their quaint Yankee names: "Clinton," "Post-Oak," "Emily." The young immigrants went cheerfully into the warming soil, and seemed to prosper.

The following summer, in a vineyard at the village of Pujaut, perched on a plateau of dry, pebbly soil a few kilometers southwest of Roquemaure, something strange started to happen. A cluster of vines began to show curious symptoms. The leaves rapidly turned yellowy, the edges reddened, the blush spread and, in an eerily accelerated autumn, by August the foliage had dried up completely and dropped. Unlike the *oïdium*, there was

43

no obvious attacker, no mildew or external signs of leaf infection.

By 1864 Borty's own Grenache and Alicante vines were shrivelling, while some native vines on a patch of sandy soil seemed mysteriously unaffected. The American imports were healthier than ever. The wine merchant's strange biological secret stayed sealed behind the high walls of his suburban vineyard.

In Pujaut meanwhile, in that second summer of affliction, everything was dead or dying. No one knew why. Winter came. The sickness abated. In the spring of 1865 the malady reawoke, descended from the plateau and extended north in a necklace of satellites beyond the town of Orange, with a southerly outpost at Saint-Rémy. That summer it crossed the river Rhône into the department of the Vaucluse. When the dead vines were dug up (no one thought to exhume a healthy one) their woody roots were blackened, and crumbled in the hand, showing a condition known as *pourridié*. This "decay" appeared to be caused by a pale fungus farmers colloquially called *le blanquet*. No other agent of attack was apparent.

Pourridié was an old problem, observed typically when vines were planted on cleared woodland. Rotting tree-roots underground hosted the ghostly mycelium. The solution seemed simple: plant vines somewhere else. But this was something different. The way the affliction struck had never been observed before. That summer of 1865 M. Barret, advocate and proprietor of Sarrians on the left bank of the Rhône, watched helplessly as the leaves of his vines "turned yellow and fell in a precocious manner." By September the "grapes were almost transparent with a pale ruby tint . . . when placed *en cuve* with grapes of good quality, the fermentation was incomplete and the wine poor."

That same summer, David de Pénanrun, a patrician customs inspector who owned a vineyard at Ville-les-Avignons a few kilometers south of Pujaut, first noticed something wrong with his vines. Leaves were turning brown and falling early. The affliction seemed to spread outwards in a circle. He began to investigate. It was in his neighbors' vineyards too. It returned the next year, and the next. In July 1867 he reported in a letter to the *Bulletin de la société d'agriculture de Vaucluse* "a new malady of the vine observed at Pujaut for the past three seasons." It seemed to "sleep in the winter and wake again each spring." De Pénanrun was curious enough to dig up a dead vine, and observed its corrupted roots were covered in fungal mycelium. He had tried spraying dying plants with "sulphur, lime, coal tar and petrol" as a cure, but without effect.

Le Journal d'agriculture pratique in Paris took up the strange story from the Vaucluse. The leaves of vines in the region were stunted and sick, its correspondent reported in autumn 1867. Some grapes set, then dried out and were dead by September. There was "no difference between young and old vines . . . no distinction between types of soil." The mystery malady "advanced almost on parallel lines, widening and enlarging each year."

The affliction spread. In increasingly anxious letters to the agricultural press, correspondents noted that the new malady affected young, vigorous vines planted on well-drained slopes rather than those on wet, heavy soil where *le blanquet* might be expected. Besides, it was spreading too fast. Some blamed winter frosts, "soil-exhaustion" or the condition known as *Rougeot* (a fungal infection that attacked vines in wet springs (this time not an unfortunate import from America). Whatever it was, more and more vines were dying.

In November 1867, a certain M. Delorme, veterinary surgeon and vineyard proprietor of Arles in the Bouches-de-Rhône, wrote to the president of the Comice Agricole of Aix that a "new disease" was developing very quickly in a vineyard planted four years earlier thirty kilometers south of Pujaut in the Crau district. La Crau was an arid, pebbly saucer of land near Arles where nothing useful would grow except the vine. The vineyards had all looked "beautiful," the harvest of 1865 had been abundant, then in July 1866 a small proprietor at Saint-Martin-de-Crau noticed leaves on a number of vines turning rapidly from green to red. Within a month "most of the vines were already withered and beginning to dry out."

Delorme had gone to investigate. "Digging one up, I found the roots as sick as anything above ground," he wrote; "they were blackened and crumbled easily into dry wood . . . The disease has not stopped extending from vine to vine in all directions – always from neighbor to neighbor – and now occupies an area of about five hectares."

Maps of the malady's advance, produced later by Professor E. Duclaux of the Lyon Faculty of Sciences and published in the reports of the Academy of Sciences, show it spreading like a slow-motion invasion that might have been conducted, in another century, by paratroops. The small Pujaut spot of 1865 has swelled a year later into the *tache de Rocquemaure*. By 1867 the area infested looks like an hour-glass, with its waist at Avignon and two elongated blooms of contiguous infection by now touching three departments – the Gard and the Vaucluse straddling both sides of the river Rhône, and its southern flank reaching into the Bouches-du-Rhône towards the sea. By 1868 the malady had marched north to Grignan in the Drôme, with a satellite spot at Bonlieue near Montelimar. It was advancing eastwards along the Durance valley in the Vaucluse, and in the Gard to the west was beginning to form a bulging salient with an outrider at Saint-Vincent. In the south it had reached the sea at Les Martigues.

The summer vineyards turned yellow, then brown. There were no grapes to harvest. Exhumed vines showed nothing except fungal rot on corrupted roots. There were no instructions from the center, no prefecturial edicts. Some bewildered *vignerons* dug up and burned their dead. Others, according to a correspondent of the *Messager agricole du Midi*, left the withered stumps in the ground "as if life would miraculously return to them with the spring rains." The plague was a judgement of God for the sins of man. Neither the village priest nor his bishop were about to condemn peasant superstition when the rampant doctrines of "positivism" threatened to raise science to the new official religion of France. Incense perfumed the vines as little processions wound among them to make their Sunday devotions and pray for deliverance. In 1868 relics of St. Valentine were installed in the church of Saint-Jean-l'Évangéliste at Roquemaure and prayers said for the saint's intercession against whatever was killing the vines.*

No miracles were forthcoming. "A shout of distress escapes from the mouths of wine growers," M. Barral, a scientifically minded journalist and vine-grower of Faugio in the Hérault, recorded. " 'The vineyard is sick, ruin is at our doors!,' they cry. Letters from stricken villages plead for an explanation – for some light of hope – for not only have the products of the vine disappeared, without them their own very existence is threatened."

Something had to be done. That summer M. Gautier, mayor of Saint-Rémy-en-Provence, and Gustave Levat, "engineer of Arles," acting on behalf of several proprietors of the Crau area, approached the Société Centrale d'Agriculture de l'Hérault for help. The Montpellier-based society was renowned for its investigations into maladies of the vine. One of its members, Henri Marès, had been a leading light in advancing the use of sulphur in the fight against the *oïdium*.

---

* It would later become the custom in the Bordelais for *vignerons* to make crosses of hazel branches, bound by wicker and garlanded with flowers, to be blessed by priests on the first Sunday of May and planted in the awakening vineyards to ward off the evil. The Catholic Church would grow alarmed as peasants invoked an older "religion" of animist naturalism to spare them from the plague.

Reports that something strange was happening in the vineyards of the Rhône valley crossed the desk of Edmond Gressier, the imperial minister of agriculture and commerce. In December 1868, "wanting an accurate account of the true state of things and anxious to make an informed judgement on the contrary opinions being advanced by vine-growers," Gressier formally requested agricultural societies in the Midi to form special investigative commissions. In the Hérault they were already on the case. The previous spring a brave little body of biological sleuths had formed, styling itself "La Commission pour Combattre la Nouvelle Maladie de la Vigne." Its members were Gaston Bazille, president of the Hérault Agricultural Society and a big landowner (he was father of the painter Frédéric), Félix Sahut, the *savant montpelliérain* and noted horticulturalist, Frédéric Cazalis, director of the *Messager agricole du Midi*, the entomologist Louis Vialla, and the forty-five-year-old Jules-Émile Planchon, professor of botany in the Faculty of Science and in the School of Pharmacy of Montpellier.

They were lucky with Planchon. He was born to a Protestant family at Ganges, thirty kilometers north of Montpellier in the uplands of the Cévennes, in 1823. His father was a candlemaker. He had been apprenticed to a pharmacist – his mornings were spent gathering and preparing plants for herbal remedies. It was a field education in botany. Through hard work and intelligence he won a place at the famous Montpellier School of Pharmacy. As a young man he had been assistant to Sir William Hooker at Kew, and taught French to the famous botanist's son Joseph, who would succeed his father as president of the Botanic Society of London. He had held important academic posts at Ghent and Nancy, returning to Montpellier in 1853 to publish works on the culture of the truffle and on the parasites of alfalfa. In the early 1860s he had corresponded (his friend Joseph Hooker was the intermediary) with Charles Darwin on aspects of hybrid infertility while the evolutionist was researching *The Variation of Animals and Plants under Domes-*

tication. Natural science was a family affair: in 1855 he had married Delia Lichtenstein, whose brother Jules was a gifted entomologist.

The little "vine malady" commission got out into the field with urgency. On the morning of 15 July 1868 the commissioners assembled armed with magnifying glasses and notebooks at the domain of the Marquis de Lagoy near the village of Saint-Rémy (curiously perhaps, the birthplace of the apocalyptic sixteenth-century seer Nostradamus) in the Bouches-du-Rhône. The day was hot, and M. Galtier, the vineyard manager, seemed agitated. What might the learned gentlemen in their top hats and somber frock-coats discover? Mourning dress was appropriate. In the center of the field a "veritable cemetery," in Sahut's description, of desiccated dead vines stood sorrowfully among the living. Bemused farm laborers set to work. Spades clanged on the dry, stony soil. Planchon published his own highly personalized account of what they found in the Montpellier newspaper *Le Messager du Midi* a week later.

"The delegates studied the affected vineyard with close attention," he wrote. "Naturally addressing the most afflicted vines first, they found rotted roots without trace of fungi or insects – circumstances that we now understand, but which at the time clouded the investigation. Nevertheless the speed of the disease's advance, its expansion from an initial center – everything indicated that the cause was living.

" 'It is terrible,' said the vineyard's manager in picturesque language, 'it advances like an army.' His words spurred us on to new searches," Planchon recorded.

A happy pickaxe blow unearthed some roots on which
I could see with the naked eye some yellowish spots. A
magnifying glass revealed them to be clumps of insects

– which on closer inspection showed a relationship with aphids and cochineal.

They were in all the degrees of their summer state, from egg to adult mother, surrounded by her numerous off-spring and probably with its descendants in the different degrees. With a little knowledge of insects one could quickly see that it belonged to the group Aphidiae *or puce-rons. But it is not a true* puceron: *it differed from the Aphis proper by . . . the fact that all the individuals observed were in the wingless state and that it continued to lay eggs during the warm season, the time when true* pucerons *are giving birth to live young.*

From this moment, a fact of capital importance was established. It was that an almost invisible insect, shying away underground and multiplying there by myriads of individuals, could bring about the exhaustion of even the strongest vine.

Telling the farmers of the Midi that some subterranean force was about to wipe out their patrimony was not easy. The argu-ments were fierce – it was the recent run of cold winters and dry summers, it was "soil exhaustion," it was anything but some horrible new unknown. The greatly respected Henri Marès, hero of the *oïdium* fight, declared it was "the severe cold that had continued unbroken last winter that is responsible for the deplorable condition of the vines – there is no evidence of any other particular malady."

In its eventual report the Hérault commission divided four to four. The Planchon faction concluded bluntly: "Kill the in-sects and you will save the plant." The Marès faction noted: "We admit that the insect is the visible cause of the malady – but there are other causes which might predispose the vine to attack – such as climate and soil. It would be a mistake to concentrate research solely on a supposed single cause." The commission of inquiry set up by vine-growers in the Vaucluse voted that an insect was the unique cause of their plight by a majority of five to four.

Planchon understood. The attacker was a tiny root-sucking aphid of a type unknown. Why had it not been observed before on dead vines? Because they were dead – the insect had moved on. Vine-growers had not thought to uproot apparently healthy vines. Those were "the circumstances that we now understand." The fortunate pickaxe blow had uprooted a "healthy" vine in the early stages of underground attack where the insects were feasting.*

The work at Saint-Rémy continued the next day, and the next. The yellow infestation was in "a hundred places, everywhere where the vines were suffering," wrote Planchon. The committee widened the search, visiting vineyards at Orange, Châteauneuf-du-Pape, Graveson and at Saint-Martin-de-Crau – whose plight the veterinary surgeon M. Delorme had reported the previous autumn. At Saint-Martin, according to Félix Sahut, "We were able to observe the first example of large-scale mortality which by now extended over eight hectares." Girdling the circle of devastation was a bigger ring of vines in the early stages of attack – with new splashes of infection breaking out at the margins into healthy plantations.

The depredation would typically begin in the center of a vineyard, spreading in a slightly elongated circle. Some *vignerons* called it *une lune*. Gaston Bazille compared the ever-expanding pattern to the way an oil spot – *une tache d'huile* – spread on water. That analogy stuck.

How did this wingless insect spread? Did it move under or above ground, or both? Professor Planchon thought it might "make its voyages only at night." If it travelled above ground, then "painting the foot of the vine with coal tar would probably

---

* Thirty-two years later Félix Sahut claimed he was the first to pull up a "healthy" plant and see the live insects, passing it to Planchon – as the botanist had privately admitted to be true before his death in 1888.

present the insect invader with an insurmountable obstacle," he wrote. But if it could move "in the depths of the soil," then stopping its progress was going to be far more difficult. The Hérault committee provided every schoolmaster in the "threatened areas" of the department with a magnifying glass and crudely-drawn instructions on what to look for. It was hopelessly amateur, but at least it spread the notion of a generalized peril. The idea of a phylloxera observation corps would be revived later by the government, with less than satisfactory results.

Planchon meanwhile could only try to answer pressing questions through plodding scientific methodology. "Where did this insect come from?" he asked. "Had it already been described? Which were its closest relatives? . . . These questions were not easy to resolve straight away, they could be answered only when the insect had been discovered and observed in all its life-stages. Having seen at first only subterranean insects devoid of wings, I looked stubbornly for a winged form, that I supposed had to exist."

The botanist consulted with his colleague C.L. Donnadieu, *préparateur zoologique* at the Montpellier Faculty of Science. After conducting microscopic dissections of the tiny insects Donnadieu declared them to be a member of the group *Aphidae*. The discovery was formally announced in a report to the Academy of Sciences in Paris on 3 August 1868 as a new genus, *Rhizaphis vastatrix* – "root aphid devastator." No male form had been identified.

At Montpellier meanwhile more members were appointed to the vine malady commission – including the youthful Camille Saintpierre, professor of the medical school and himself the owner of a large vineyard at Rochet. Driven, in Planchon's words, by "agitation on all sides, by numerous pleading letters from the agricultural population demanding that our experiments, our hopes and our failures be confided to them," on 23 July 1868 he and Saintpierre began a week of experiments at Rochet to see what might kill the "dangerous aphid." It was all pretty desperate. Scientific precepts learned during the battle against the *oïdium* were mixed up with gleanings from horticultural folklore. They drenched vine footings in diluted petroleum, heavy oil, coal tar, "black soap," arsenical acid, phenic acid, salts of copper, "oil of cade," infusions of tobacco and mustard, aloes and walnut leaves. Whatever damage these supposed remedies might do to the vines themselves, in their deep root fastness the insects sucked on happily.

How did the insect survive the winter to awake in the spring? Planchon and Saintpierre admitted they did not know – perhaps it was as an egg. On 12 December 1868 they found a cluster of tiny larval young, "more or less numb," clustered on a root. They seemed to be hibernating.

9

🦗

There was a bizarre interlude at the Mas Fabre vineyard near Gravéson, owned by an enlightened *viticulteur* called Louis Faucon, who had read Planchon's theories published in the newspapers with consuming interest. That summer of 1868 Faucon's vines began to sicken. He posted his two young nephews to watch over them. One late July morning they saw microscopic yellow specks moving across the surface of the sun-baked soil. Planchon reported what happened next: "These spirited youngsters told their uncle that they had seen the insects strolling along like good bourgeois going into a restaurant with walking sticks in their hands."

The proprietor had repeated this curious observation to the reporter from the *Messager agricole*. The story ran on 5 August. "M. Faucon made the very excusable mistake of repeating the expressions of his imaginative nephews," wrote Planchon. "A lot of people did not believe that the insects walked like people with canes in their hand."

Of course they didn't, but in the febrile atmosphere of the Midi that summer, anything was believable. Just why the practical Faucon should use such colorful language remains a mystery. Perhaps he was caught up in the general hysteria. In any case, the discovery by his nephews that the insect could move across the ground was discounted.

It was the still-elusive winged form that most perplexed Planchon, however. He had been diligently scraping yellow clusters from infected roots into a "hermetically sealed" collecting jar from the very first day. At the vineyard of M. le Marquis de Lagoy on 15 July 1868 he had chanced on a tiny pale-yellow "nymph" with, as he described it, "wing-buds on two sides of its body like triangular tongues." The insect had died. He captured more wing-budded nymphs at vineyards near Sorgues and Gigondas in the Vaucluse on 17–18 August. He carefully corralled them in a laboratory jar and waited.

Planchon meanwhile sent field notes and live samples of the new discoveries to Professor Victor Signoret in Paris, foremost entomologist of France (author of the *Hemiptera* chapter in the hugely popular *Guide de l'amateur d'insectes* among other things). If anyone might know more about the aphid it was the famous *hémipteriste*. The Parisian *savant* noted a correlation of the wing-budded nymph with a known pest of oak trees which had been named as a species in the 1830s by the Provençal naturalist Boyer de Fonscolombe. He had called it *Phylloxera quercus*.

The troublesome oak aphid attacked leaves directly, not the roots of trees. It showed both winged and wingless female forms, but the rest of its life-cycle remained mysterious. Its pricking depredations however had a similar desiccating effect on foliage to whatever was attacking vines. Professor Signoret suggested replacing the root-derived designation *Rhizaphis* of the insect Planchon had discovered with *Phylloxera* – Greek for "dry leaf." The provincial botanist deferred to the great Parisian insect expert. The villain had its name: *Phylloxera vastatrix* – dry-leaf devastator.

In Montpellier meanwhile, one at least of Planchon's captive wing-budded nymphs was moving to her last larval moult. "This form indeed existed," he wrote, "and having discovered it as a nymph with its wings still enclosed by their sheaths, on 28 August 1868 I observed it hatching into an elegant little *moucheron*,* looking like a cicada in miniature, bearing four flat, transparent wings. From then on my rhizaphis became a phylloxera."

The new designation was first announced in the *Messager du Midi* on 31 August 1868. Now the assassin at least had a name. Planchon posted a more formal notice to the Academy of Sciences in Paris two weeks later. His report described his experiments with more living specimens. How did they move? How did they find food? How did they spread so fast?

"In a box one meter long I put some fresh earth taken from a garden in Montpellier that was free of pests," Planchon wrote.

> *In this earth, I carefully placed some sections of vine which were infested with the wingless aphids.*
>
> *I covered each section with a glass bell-jar slightly raised on one side to allow the insects to get out. I placed fragments of healthy vine-roots about three centimetres away . . .*
>
> *The next morning three young aphids had colonised the nearest fragment . . . some hours later there were twenty young on it.*

Planchon described the antics of his aphid circus with paternal affection. "In the first days of their active life, the young are *à l'état vagabond*," he observed. "They wander here and there looking for a suitable place to fix themselves. They move faster than those in the adult state, they seem to palpate and sense their surroundings with their antennae, if you will forgive the comparison, as a blind man explores the ground with his stick before venturing further." With their "eyes" apparently nothing more than primitive patches of pigment, Planchon assumed the insects found new feeding grounds by some sense of smell.

He had managed to hatch out more winged forms. Under

---

* "Midge" – "in a vague and non-scientific sense of the word," Planchon added.

observation in their sealed jars his captive aeronauts seemed to prefer sitting pretty still. They laid eggs and died. The eggs remained inert. He assumed this was to do with the conditions of captivity, and thought that airborne dispersal by these winged, egg-bearing females was the principal agent of the malady's long-distance spread. It was the wind of Provence, the Mistral, which had disseminated the insect invader the length of the lower Rhône valley, he proposed. But there was hope. The prevailing north-west wind of lower Languedoc must, he informed the Academy of Sciences, return them to their original center of propagation. The Marin, the coastal wind that also blew from east to west towards Montpellier, was habitually charged with rain, which would wash out any aphids. He was wrong.

The disaster beginning to unfold in the Midi went unnoticed in Paris. The capital was gripped by the imperial government's jerky moves to a new political and economic liberalism. There was a grand military parade on 13 August, centenary of the birth of Napoleon I (the Emperor Louis-Napoleon was too ill to attend). Old soldiers of the *Grande Armée* raised a glass or two to celebrate the announcement of an increase in their pensions. The grape harvest across the rest of France looked as if it would be especially abundant. The good weather held. In the Bordelais an excellent vintage was predicted.

Jules Lichtenstein gave a brief lecture to the members of the Société d'Entomologie in the capital featuring a *préparation microscopique* of the aphid – and later that summer of 1868 Jean Tapié, editor of *Le Petit journal* and keen *viticulteur*, took up the curious story about strange insects and dying vines in the lower Rhône valley, publishing some of Planchon's observations in his robustly populist newspaper. One correspondent claimed that because aphids secreted a honey-like substance, hungry ants would surely polish them off. Dr. Jules Guyot argued in the *Moniteur vinicole* that the "new malady" had been around for years, and was a condition known in some wine-growing regions as *Cottis* or *pousse en ortille*. It was the result, he insisted, of "over-pruning." "Why should we declare our vineyards to the outside world as being the receptacles of every infirmity?" he wrote. "Why make such a fuss over so little?" The metropolis was otherwise uninterested.

In autumnal Oxford, meanwhile, Professor Westwood was already making connections. Could the "new" root-sucking aphid identified in France be the same as the one he had been sent from Hammersmith five years earlier enclosed in its leaf-gall "tomb"? He had been sent the samples of blistered leaves from Wicklow and Cheshire. Then in October the year before, the first evidence had come from the Cheshire greenhouse of the underground form. The insects seemed to be identical. The professor wrote:

> *In the autumn of 1867, and during the past year my attention has . . . been several times directed to the same insect, which appears to have become extensively disseminated, and has exhibited its powers of mischief in a most unlooked-for manner – since not only have I received further specimens of the vine leaves infested [by galls] in the manner above-mentioned, but have had portions of the roots of vines sent to me from different quarters, the rootlets of which had been sucked by a wingless insect, which I cannot in any manner distinguish from those of the galls on the leaves.*

On 30 January 1869 Professor Westwood at last published his observations in the *Gardener's Chronicle and Agricultural Gazette*, telling the story of the Hammersmith discovery of six years earlier, the more recent discoveries of both leaf-gall and root-infesting forms of the aphid, and how he had "communicated a notice" the year before to the Ashmolean Society, naming the curious insect *Peritymbia vitisana*. It was illustrated by a woodcut of the insect and a gall-encrusted leaf. He summarized recent developments:

> *In France, where the culture of the vine is of much more popular importance than in England, the disease has manifested itself with great virulence. At the meeting of the Entomological Society of France on the 12th of last August, M. Lichtenstein communicated a notice of the ravages of the insect, which was stated to destroy the vines only on the left bank of the Rhine [sic], from Arles to Orange, together with a notice of M. Planchon's observations, and with the remark that M. Signoret, the distinguished entomologist of Paris (whose attention has for some years past been devoted to the Coccidae and allied insects), considered that the insect belongs to the genus Phylloxera.*

Westwood referred to Planchon's discovery of the underground form, and the Montpellier botanist's name for it, *Rhizaphis vastatrix* – root aphid: "A name, as 'M.J.B.' [the Reverend Miles Joseph Berkeley] well observes, [seems] scarcely applicable, should it turn out, as we suspect will be the case, to be congeneric with the very similar insect which is found in the excrescences on vine leaves." Were they the same? Professor Planchon was about to find out for himself.

10

The vineyards of the Midi fell into winter sleep. In Paris three Ministers of Agriculture came and went in the space of twelve months. When the aphids awoke again in the spring of 1869 the government still proved languidly indifferent to what might or might not be happening on the flanks of the Rhône valley. It fell to the Société des Agriculteurs de France to sponsor a renewed inquiry, sending Vialla, Planchon and Lichtenstein back into the field. On 11 July the brothers-in-law were examining the vineyard of Henri Leenhardt near the village of Sorgues in the Vaucluse.

"On the footings of some vines of a variety called 'Tinto'* we observed some red excrescences standing out from the deep green of the leaves," Planchon recorded. "In the cavities of these galls we found phylloxera in varying numbers, typically in these combinations, in the first case one or two egg-laying mothers without wings, some of them already dead, in the second case a small number of young and some eggs." But when the mysterious "Tinto" was exhumed from the ground, its roots were healthy. This was most curious. There was something even stranger. "Observing mothers, young and eggs we could see no essential difference to those phylloxera we had

---

* Just what species or variety "Tinto" was, Planchon was unsure. Whether it was European or American would prove significant later. According to the French ampelographer Pierre Galet, "Tinto" is a local Vaucluse name for the vinifera variety Grenache.

already found on roots," Planchon wrote. "When we got back to Montpellier our leaf-galls were dried out, the insects had also suffered; a precise comparison between them became difficult, nevertheless we sent the Entomological Society of Paris a short note of our findings."

The investigators had little time to relax. A few days later a summons arrived from Bordeaux. Three hundred kilometers to the west of Montpellier something strange was happening in the vineyard of M. Chaigneau, doctor of medicine and proprietor of the domain of Gravettes in the commune of Floirac. His vines were dying. Planchon and Vialla arrived on 16 July, magnifying glasses and collecting jars in hand. The doctor was bemused. When his vines were uprooted they were covered in ugly "nodosities" and a mass of sucking aphids. Planchon found a winged specimen. Dr. Chaigneau told them a now familiar story – the first signs that something was wrong had appeared three years earlier. The affliction had awoken the following year with new voracity. Now, in the summer of 1869, ten hectares were infected and a satellite outbreak had been reported at Saint-Loubès fifteen kilometers to the north. The physician had descriptively called the malady the *Phthisie galopante*.

Where it had come from was utterly mysterious.

The answer was not long in coming. Not far from Floirac was "La Touratte," the estate of Léo Laliman. He had been a connoisseur of exotic vines for decades, written books in their praise and had eagerly imported samples from America. Like his neighbor Dr. Chaigneau he had noticed a little spot of dying vines two years before. "I had cultivated American vines for about twenty years in all innocence," he wrote later. "When the disease came my [American] Labruscas were at first untouched but the plague suddenly appeared on my [European] Alicante and on my Grenache."

That July, Laliman had discovered leaf-galls on some of his

beloved Vitis cordifolia that had been imported two years previously from Mr. J.P. Berckmann's "Fruitland" nursery at Augusta, Georgia. The ugly galls were also apparent on a Malbec (a European vine) "of which three branches interleaved with the American variety." He found two forms of tiny insect inside – one "fat and torpid," the other "small and agile." On 30 July 1869 Laliman carefully packed the blistered foliage and their living inhabitants into two parcels, sending one to Professor Signoret in Paris and the other to Planchon for comment. It was the beginning of a stormy relationship.

The root-sucking and leaf-gall forms were the same. But how to prove it? On 6 August 1869 Planchon and Lichtenstein shut the gall-dwellers conveniently sent from Bordeaux in a glass tube with a supply of vine leaves and root from the same plant. For some reason the insects stopped pricking the fresh leaves. They appeared to be "in a state of migration." Many died on the glass walls of their prison, rapidly turning a "cadaverous black," but a "certain number attached themselves with a sort of predilection to the roots" and began to feed greedily.*

Jules Lichtenstein suspected something more. This leaf-living aphid had been described before. Notice of an insect that formed galls on the leaves of vines had been in the literature for well over a decade. It had been published in 1856, far from the Rhône valley, in the *Annual Report of the Agricultural Society of New York*. In a chapter on insects harmful to fruit trees, Dr. Asa Fitch, the state entomologist, had described "globular galls the size of a pea on the leaves of the vine in which we find a pale yellow aphid." The infestation was not fatal. There was nothing to indicate that the curious insect, whose actions were similar to an aphid that caused leaf-galls on elm and poplar trees, was in any way dangerous. There was no mention of a root-living form.

---

* Reversing this experiment, turning root-suckers into "gallicoles," would prove much more difficult – although it can happen under certain conditions in nature.

Dr. Fitch had called his discovery *Pemphigus vitifolii*. Lichtenstein remembered the short mention. Could the insects found at Sorgues and La Touratte be the same as that described in New York State thirteen years before? He was convinced they were, and published his theory in the *Messager agricole du Midi* on 5 September 1869. Somehow the insect had come from America. On Lichtenstein's urging, Professor Signoret consulted the whiskery New York report, put it together with the Vaucluse discovery of red-galled leaves, and concluded broadly that the description of the American aphid matched that of the species that the brothers-in-law had found.

The professor conceded just a little to Lichtenstein: "It is not impossible perhaps," he wrote, "that the insect or its eggs had been imported with American vines . . ."

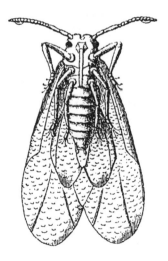

11

If the delinquent aphid had indeed come from America, an (English-born) American was already in hot pursuit. In 1866 the Horticulture Society of Missouri had "memorialized" the legislature to appoint a state entomologist. New York had such a scientific ornament, so did Illinois. Why should the great agricultural powerhouse of the Mississippi valley not have the same? There was a suitable candidate, a young Englishman working as a journalist in Chicago, currently engaged in publishing the journal *American Entomologist*. His illustrations therein were thought of great merit. The gifted insect-artist had been eagerly sketching beetles since childhood.

Charles Valentine Riley was born in London in 1843. At a tender age he showed "a marked taste for insects," and was evidently encouraged in his pursuits by a family friend, a certain "Mr Hewitson of Weybridge who had a large collection of insects and birds." Schools in Dieppe and Bonn followed. At the age of seventeen he departed Europe on an emigrant ship. He worked on a cattle farm in Illinois, became seriously ill, then recovered to get a job in Chicago writing on insects for the *Prairie Farmer* newspaper. The Colorado beetle was on the march, and so was the Confederacy. In 1864 Riley's newspaper career was interrupted by six months' service as a private in the 134th Illinois Volunteers. With the victory of the Union, Riley returned to his insect-hunting and began, among many other things, investigations into a tiny aphid that caused curious galls to form on the leaves of American vines.

The Missouri State Board of Agriculture announced Riley's appointment in 1868. From then on he prepared a total of nine successive annual reports on "beneficial and noxious insects." They were written in a homily-studded style, with comic "Darkies" and "Cockneys" from his London youth offering homespun observations on the antics of the insect world. Darwin greatly admired them.

Investigating a parasite of the vine was no dry scientific puzzle. They made wine in Missouri, a million gallons of it in the year of Riley's appointment – a third of Ohio's production and a seventh of California's, but of growing economic importance nevertheless. From Missouri's vineyards, eventually, would come redemption for the vineyards of France. But that the vine was cultivated at all in the state sprang from a Germanic root.

In 1824 a group of German immigrants in Philadelphia had formed the American Association for the Promotion of German Settlements in Western States. After several false starts they acquired twelve thousand acres of wilderness west of St. Louis along the Missouri River. A little town, Hermann, sprang up, a model Teutonic settlement with neat brick houses and a Lutheran church. The soil and climate were benign, and wild grapes grew abundantly, but no attempt had been made to cultivate them. Hermann's first settlers, clerks and advocates before they were farmers, went sadly wineless.

In 1838 Martin Husmann, a village schoolmaster from Brandenburg, arrived at Hermann with his wife and three children. He acquired a few Isabella vines. A small crop of grapes was produced in 1845, and a little wine was made the following year. It seems to have been palatable enough. In 1849 his son George headed west, like so many others, lured by the California goldfields. He returned to Missouri on his father's death, making the journey in a Conestoga wagon bearing an assortment of vine cuttings, to tend and expand the fledgling plots.

"I well remember the first cultivated grape-vine which produced wine in Hermann," wrote George Husmann twenty years later. "It was an Isabella, planted by Mr Fugger on the corner of Main and Schiller streets and trained over an arbor. So plentifully did it bear that several persons were encouraged by this apparent success to plant vines."

The "success" was illusory. The hopeful little plantations of Isabella and Catawba vines succumbed to rot and mildew. The wine they made was not up to much. Husmann found salvation meanwhile in Norton's Virginia and Concord grapes – the wine from which, he wrote in 1865, "is evidently destined to become one of the common drinks of our laboring classes." His vision was now unbounded. The Union's victory had made "all Americans truly free," he wrote at the end of the War between the States. As demobilized soldiers headed home they should "turn their swords into plough-shares and pruning hooks" to build the great republic anew as an oenological utopia, where drinking wine would abolish the curse of drunkenness and promote good will to all men.

"Truly the future lies before us rich in glorious promise," Husmann wrote ". . . and ere long . . . America will be from the Atlantic to the Pacific one smiling and happy Wineland; where each laborer shall sit under his own vine and none will be too poor to enjoy the purest and most wholesome of all stimulants, good, cheap, native wine."

The long journey of Isidor Bush from Bohemia to the banks of the Mississippi was an eventful one, and his vision for America's wine industry was perhaps more commercial than George Husmann's. He was born in 1822, the son of a wealthy Jewish merchant. Political journalism got him into trouble with the police and he fled Habsburg-ruled Prague for New York in 1849, where he began publishing *Israel's Herald*, the first Jewish weekly newspaper in the United States. When the business failed he moved to St. Louis. In 1861 he became the city's representative to the Constitutional State Convention that deposed the Confederacy-sympathizing governor and kept Missouri in the Union. The entrepreneurial Mr. Bush acted as agent for the Iron Mountain Railroad through the upheavals of the Civil War, then in 1866 acquired 240 acres of land below St. Louis along the Mississippi River.

He turned to viticulture, and his nursery at "Bushberg" became the largest in North America. Bush & Sons' illustrated catalogue, first published in 1869, was a mini-masterpiece of American ampelography. The same year Isidor Bush ambitiously founded the Bluffton Wine Company with a model township, a profit-sharing scheme for small farmers and George Husmann as president. This utopian project went bust in 1871. The nursery business at Bushberg, however, was to prosper for reasons no one could have foreseen.

George Husmann set up his own plant nursery at Sedalia, Missouri. In 1867 a Swiss-German immigrant named Hermann Jaeger planted a vineyard and nursery at New Switzerland in Newton County. Messrs. Poeschel and Scherer (later the Stone Hill Wine Company) opened the biggest of Hermann's wineries in 1869. The state of Missouri had become a vine factory looking for customers. Although he did not know it yet, Charles Valentine Riley would make the necessary introductions.

In the early summer of 1869 Charles Riley read Professor Westwood's tale of a mysterious insect found infesting the leaves and roots of vines in British greenhouses – the same agent apparently responsible for a new malady of the vine reported in agricultural journals to be afflicting south-east France. He had himself encountered such an insect in America, and had published a report on a "grape-leaf gall-louse" in the *Prairie Farmer* in late 1866. Examining a tiny aphid found in leaf-galls on plantations of Clintons and Taylors, Riley had proposed, as Signoret would do three years later, that it was a species of the known oak-tree pest *Phylloxera quercus*, that had long been in the literature.

The vine-leaf insect had been described more fully in 1867 by the Philadelphia naturalist Dr. Henry Schimer. He had opened "thousands of galls" yet found only a very small number of winged specimens. He assumed them (wrongly) to be the male of the species, and had concluded that it was an aphid belonging to a new family which he proposed should be called the *Daktulosphaira*. Riley's colleague Benjamin Walsh, state entomologist of Illinois, also described it in the autumn of 1867, and had given it yet another name – *Viteus vitifolii*. None of these learned gentlemen had ever observed a root-living form.

That autumn of 1869, Charles Riley made contact with Lichtenstein and Signoret in Paris. He sent desiccated leaves containing the gall-living insects across the Atlantic, accompanied by a résumé of the American discoveries. With his report to the Entomological Society of France about to go to press, Professor Signoret admitted he "had not had time to peruse these findings and would make no judgement on just what the insect should be called until I have done so."

The French Société des Agriculteurs meanwhile had published a collection of highly descriptive reports gathered by their inquiry commission from the front line in the Rhône valley. At Pujaut, the investigative Louis Vialla found "all the vines growing on the stony plateau had perished – those growing on other types of soil had been attacked later – but all had succumbed." In the town square of Roquemaure *vignerons* were selling their ruined vines as firewood at eighty centimes per hundred kilos.

M. Marin, the mayor, shepherded the investigators through the little communes around Roquemaure – Saint-Geniez, Tavel, Saint-Laurent-des-Arbres. "The aphid was everywhere – where one took the trouble to look for it." There was an interesting new observation along the way. Vines growing on a bank of sandy soil at Louis Faucon's vineyard at Gravéson in the Bouches-du-Rhône had stayed healthy, while all around them his Grenache had succumbed. The reports were first published in the *Journal d'agriculture pratique* and then as a booklet with an analysis by Planchon of what he knew of the insect's life-cycle inserted as an addendum. That the insect was the cause of the new vine malady he for one had no doubt, although the committee's *rapporteur*, Louis Vialla, had to admit that there were many who still thought the opposite.

In Paris, Professor Victor Signoret read the Planchon-Vialla reports grumpily. To him, the detailed descriptions from where "the malady reigned" of soil quality, climatic history and methods of cultivation merely proved that his own theory was right. Oddly perhaps for an entomologist, he blamed everything but the tiny aphid. While he conceded that a large quantity of root-sucking insects might cause the plant "disadvantage," the real reason, he insisted, was either drought, poor soil quality or "bad culture." The climate was to blame – a run of exceptionally dry summers had caused rotting of the roots. "It is this decay which causes the death of the vine and not the aphid, which leaves the root well before it is dried out," said Signoret.* The predations of the opportunist insect were a symptom of some other malaise. France's wine capital sided with the Parisian professor. The Société Linnéenne de Bordeaux published its first report on the "new vine malady" in August 1869, promising to take no sides in the cause-or-effect debate. By December however the prestigious society's secretary, the entomologist A.-H. Trimoulet, was convinced. The aphid "was not the direct

---

* The insect abandoned its feeding ground when the sick vine's root-sap pressure dropped, and it went in search of new ones. That is why phylloxera was not observed on dead vines.

cause" of the affliction, he pronounced. It was impoverishment of the soil due to injudicious irrigation schemes. Trimoulet could not attack the false doctrines of Planchon's *système phylloxérique* robustly enough. Where was the proof the insect had somehow come from America? He would continue to say the same thing for years.

That the root-sucking insect was the cause of the malady was obvious to the *phylloxéristes* of Montpellier. That the insects found above and below ground were the same was also increasingly beyond doubt. That they had somehow been imported from America, as Jules Lichtenstein was now controversially suggesting, was a possibility. Planchon himself was not yet certain. "The phylloxera may have existed already for a long time in our country and have lived here unnoticed and harmless until the day when it abruptly multiplied because of new circumstances," he wrote in December 1869. "This idea has many supporters . . . on the other hand the insect might have arrived from some region unknown, but this last supposition has no solid foundations."

That November the French Société des Agriculteurs held their viticultural congress at Beaune, the ancient wine capital of Burgundy. After much mutual self-congratulation the second day was given over to discussion of the "terrible malady" afflicting the vineyards of the south. Planchon was there to give a brief lecture on the horrible aphid. From information he had received, he told the delegates, it was possible there was a natural predator of the phylloxera, a native French insect that might defeat the plague. This line of research was proving hopeful. The Société gave him a medal for his efforts. Léo Laliman turned up to cheerfully inform his audience how his precious American vines had survived the onslaught of the phylloxera to varying degrees.

To the grand proprietors of the Côte d'Or the aphid might as well have been on the moon. The irrepressible Laliman however insisted on showing off his healthy specimens (he had brought them with him in dainty pots) of Taylor, solonis, York-Madeira and Jacquez. His hosts were bemused. Were they supposed to replant the noblest vineyards in France with these vulgar intruders? The report of 13 November 1869 in the *Messager du Midi*, however, was detailed and animated. Laliman had mentioned using these varieties as "rootstocks" on which vinifera might be grafted. From his own experiences he had discarded "the race Labrusca" (Concord, Isabella, Catawba, etc.) as not being resistant enough. Such observations seemed the babblings of a lunatic. They would have enormous importance.

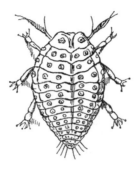

12

Cracking the devastator's life-cycle and its interaction with its host would take many years of scientific detective work. Planchon, Lichtenstein and Riley had opened the international scientific conversation. Many more would add their voices. In 1872 Maxime Cornu, professor at the Paris Museum of Natural History, began an intimate investigation of how leaf-galls were formed and just how the root-living form made its attacks. Professor Georges Balbiani of the Academy of France simultaneously embarked on a decade-long examination of the aphid's sex-life. The research effort spread to Italy, Spain, Switzerland, Austria-Hungary and Germany – wherever culture of the vine was economically important. It continues just as energetically today in California and Australia.

From 1869 onwards, as the aphid began its twenty-year march across Europe, the trail of biological discovery would be played out in the French intellectual imagination like a magazine-serialized *roman policier*. The next installment – published in the august reports of the Academy of Sciences or the columns of *Le Journal d'agriculture pratique* – was always eagerly awaited. There were plot twists and red herrings galore. A modern summation of the insect's baroque life-story demonstrates just what its nineteenth-century would-be-biographers were up against.

Daktulosphaira vitifoliae (Fitch) is a member of the order *Hemiptera* (half-winged), sucking insects that include cicadas, scale insects and the family *Aphididae*. Aphids generally can reproduce sexually or parthenogenetically (without mating and gene transfer). They can be winged (alates) or wingless (apterous). Some species of aphid require different host plants for progressive stages of their life-cycle.

Phylloxera has a single host and a single source of nutrition, the vine, but its life-cycle can vary dramatically according to which species of Vitis it makes its home. Its rate of development can further be affected by numerous environmental factors, including soil and air temperature and humidity.

The life-cycle displayed on American species is dauntingly complex. To begin at the beginning – or at least with the egg. The ticking bomb is a tiny brown speck, attached by a clever little hook to the bark of a young vine in a somnolent winter vineyard. The desiccated corpse of the insect that laid it might be nearby. In early spring, with rising temperatures this so- called winter egg lightens to a pale amber – then hatches into a yellow-colored wingless female nymph: the "fundatrix." She climbs onto the top of a young leaf, inserts her proboscis and starts to suck sap. A small depression is formed into which she slowly sinks, injecting as she feeds a vine-cell growth-altering hormone which causes the leaf to develop the characteristic gall, bulging beneath and now closing above her, leaving a bristle-guarded slot on the leaf's upper surface.

Within her little pocket, protected from the rising heat of early summer and the attentions of hungry birds and insect predators, the nymph moults several times to become an adult. She reproduces asexually, laying as many as five hundred lemon-yellow eggs inside this primary gall – within which she herself stays entombed before dying, exhausted after her oviparous epic. The eggs hatch into "gallicole" nymphs, or "crawlers," which depart the original gall through the bristle gate, move energetically to find new young leaves and themselves start to suck and sink into the upper surface – producing secondary galls of their own. The galls appear therefore like a mass of blisters on an infected leaf's underside. This second generation become gall-living adults which in turn lay eggs. A growing individual will moult four times over several weeks to become an adult, when she in turn will start laying eggs, and may live for a month or more. Up to five generations (less in cooler climates) of gall-inhabitants may be produced in a season, each progressively showing the slightly different physical characteristics of the root-living form.

Planchon was fascinated by the mathematics of the aphid's life-cycle. "As for the number of eggs that a single female can produce it varies according to the circumstances," he told a conference of farmers in June 1873. "In the crushed body of a mother on the point of laying, we have seen an ovary with twenty-seven eggs in various degrees of evolution. Thirty eggs are the most we have seen laid by a female from 15 to 24 August 1868. This gives an average of five eggs a day in a warm season.

"Taking twenty as the average number of eggs and eight as the number of layings between 15 March and 15 October, one can calculate the progressive number of individuals originating from a single female: in March, twenty; in April, four hundred; in May, eight thousand, in June, 160,000; in July, 3,200,000; in August, sixty-four million; in September, 1,280,000,000; in October, 25,600,000,000."*

In early midsummer, migrant nymphs from the above-ground population begin to move down, some crawling, others falling, to the footing of the vine, where they venture under-ground to become root-living adults – myriads of them – as

---

* Examining leaf-galls in 1871, the American entomologist C.V. Riley stated: "Small as the animal is, the product of a single year would encircle the earth thirty times in an endless line." George Ordish pointed out that at 0.1 grams per aphid, this biomass would weigh almost five metric tons, and assuming a dispersal of five fundatrices per hectare, would deliver a far bigger harvest of insects than of grapes. The vast proportion fail to find a sustaining vine leaf or are gobbled up by opportunist predators.

many as one thousand insects per ounce of root near the surface. Any adults, nymphs or eggs left behind in the galls die in the autumn when the leaves fall.

Meanwhile the migrant root-dwellers, distinguished by rows of brown tubercules on their backs, go to work on the roots, inserting their highly adapted feeding tubes into the woody tissue. They in turn lay root-eggs which hatch into root-living nymphs, which in turn become adults which lay more eggs. As the autumn soil cools, these hatch into either wing-budded nymphs or small, darker-colored hibernant nymphs which will survive the winter in suspended animation deep underground.

In the spring, when soil temperatures go above a critical level (about 45 to 65°F) and vine sap starts to flow, these hibernating nymphs wake, begin feeding and moult four times to adulthood. Newly hatched nymphs can leave the roots and travel on the soil surface, move underground in cracks in the earth (especially in claycy soils), or can climb the vine and be windblown for considerable distances. These travellers usually leave a vine only once phylloxera populations become high and there is feeding competition, or when a vine is near death.

On some of these nymphs, wing-buds develop. They become winged, egg-laying adults, greenish-yellow in color, and start to emerge from the ground in late June, to clumsily flutter short distances through the warm afternoon air. There are two flying forms, both female, which each bear two or sometimes more eggs. The smaller form, the "androphore," carries male eggs, the slightly larger "gynephore" bears female eggs. These are laid on the underside of a vine leaf.

These eggs hatch out as nymphs which go through their moults very rapidly to become wingless, sexual adults. They have no digestive organs. Their only function is to mate, which they do eagerly. The abdominal cavity of the fertilized female becomes massively engorged – she is "an egg mounted on six legs with two antennae," according to one nineteenth-century description.

She has one last task, to move down the vine to lay her single "winter egg" in a convenient bark crevice in the two- or three-year-old woody stem of a vine. Then she dies. Her mouthless, anusless male partner may proceed on his single-minded sexual

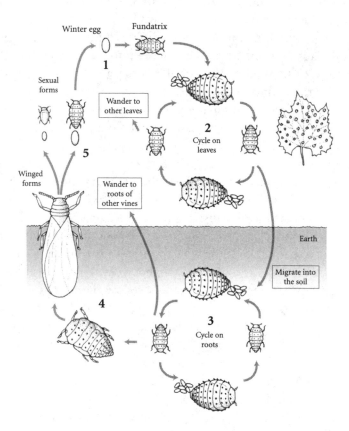

The complex life-cycle of the phylloxera took many years for scientists to fully understand. On American vine species there are both above- and below-ground phases with the intervention of winged and male and female sexual forms. On the European vine, Vitis vinifera, the insect's life is conducted almost exclusively underground and reproduction is asexual.

mission in search of new conquests for a few days yet. The cycle is complete.

As an example of parasite adaption to host, the phylloxera's multiple quick-change act is one of the most remarkable in nature. The underground stage and the winter egg provide protection through hot dry summers and very cold winters. And as a clever parasite, in its native North American range at least, its attack does not kill the host. Leaf-galling is not fatal to the vine; nor, on American species, are the root predations. Over millennia of evolution wild vines developed ways to keep the attacker at bay − a corky layer of cells that plugs a feeding site, starving the insect of nutrients; and offensive biochemical mechanisms that repel or kill crawlers.

European vine-roots had and have no such defenses. The aphids fell on the feast, rapidly dispensing with the aerial opera in a remarkable display of adaptive power. In the dry summer environment of southern Europe there was no need for winged forms, no need even for males. The life-cycle of the phylloxera on vinifera vines is almost entirely subterranean.*

The winter egg is extremely rare − as are the leaf-gall stages − although it can occur. The female-only aphids remain below ground throughout reproducing asexually. The adults lay eggs throughout their lives, sometimes as many as ten per day, which then hatch in less than a week and, as in the American cycle, become "crawlers" which move along roots and can also climb vine trunks and be blown considerable distances on the wind. They use their antennae to search for a new feeding site, where they will stay for the rest of their lives. A tiny proportion will

---

* In 1973 Pierre Galet observed phylloxera on the foliage of Vitis caribaea in the equatorial jungles of Venezuela, where there is no autumn leaf fall. He could find no root-living form. He proposed that: "By leaving the tropics for moderate climates, either by natural migration or by the intervention of man, the phylloxera was confronted with the problem of annual leaf-fall and more rigorous winters. Adaptation to a subterranean life to survive was the result."

overwinter as root hibernant nymphs which will awake in the warming spring to repeat the cycle.

What does the damage to European vines? Twenty-first-century research is still looking for the complete answer. The aphid's feeding tube is highly adapted. It has been described as "a hypodermic syringe that works in both the in- and out-directions." The needle-like mouth punctures through the root bark and into cells. Once the mouth-needle is inserted, the phylloxera confirms by taste that the root is a good source of food, then she injects saliva. This activates the root cells to increase in size and number and mobilize the stored nutrients, so that starches turn into sugars and flow towards the feeding site. The phylloxera suck up the liquid nutrients and grow.

Tapping into the plant's vascular system is not enough to kill it. In midsummer there may, as has been noted, be a thousand individual phylloxera per ounce of root near the surface. But even at this density, as the grapes begin to ripen the microscopic phylloxera will lose in the competition for sugars, because their weight is a thousand times less than the weight of the grapes pulling in the other direction.

Something more is killing the vine. The aphids generally feed on the tips of the rootlets. The saliva injections cause yellowish-brown swellings or galls (nodosities) to form which may curve and bulge around the insect itself. In most cases the swelling stops rootlet growth and the infested portion eventually dies. Feeding on larger roots causes rounded swellings (tuberosities) which give the root a "warty" appearance.

These injuries and malformations impair the vine's absorption of nutrients and water, contributing to its decline. The decomposition of roots is also hastened by secondary infection by fungi and by the feeding of other insects and mites. These opportunist invaders are the true assassins. It is thought therefore that as well as evolving resistance to direct attack by phylloxera, American species had also found ways of defeating the fellow-travelling pathogenic fungi that sought to exploit the feeding wounds.

After research in the 1990s, the California scientist Jeffrey Granett, professor of entomology at the University of California, Davis, put it bluntly: "Death is caused by plant pathogenic fungi that enter the feeding wounds of phylloxera. The fungi can be of many types and are ubiquitous in agricultural soils. When pathogens . . . are introduced into feeding wounds, they will first rot the cells that phylloxera are feeding from and then slowly girdle the root. The girdling will disrupt the continuity of the root's phloem and kill the root. If enough roots are killed, the vine will die."

Far from the now sleeping vineyards of the south, in Paris in the winter of 1869 it was remarkable just how much could be deduced of the life-history of the phylloxera from simple observation. Armed with field notes and samples from Planchon and Léo Laliman's collecting flasks, Professor Victor Signoret was able to put together an admirably comprehensive if still incomplete description of the insect's anatomy and life-cycle.

In the same learned paper in which he dismissed the aphid as the "pretended cause of the new malady of the vine" the professor nevertheless described what was known of the insect in its various life-stages in intimate detail. He admitted what he did not know. Although everything pointed to the existence of a male form of the *dévastrice*, it had yet to be found. Were the leaf-galls formed by winged or wingless females? How were the leaf-galls formed? The winged females evidently produced only a few eggs – where were they laid? And in spite of their apparent experimental interchangeability, was the gall-living aphid really the same as the root-living form? The overwhelming question – where had it come from? – had yet to be answered.

* * *

The imperial government stirred. The howls of distress from the Midi, as well as the threat to state tax revenues should the vast wine culture of Languedoc and Roussillon collapse, were too great to ignore. Committees of farmers were not sufficient to the task.

In July 1870 the Ministry of Agriculture and Commerce appointed a Commission on the Phylloxera (sometimes called the "Central Commission") chaired by the eminent chemist Professor Jean-Baptiste Dumas, permanent secretary of the Academy of Sciences, in effect chief scientist of France. Civil servants and politicians as well as scientists were impaneled to bring centralized Parisian order to the struggle. Louis Vialla and Henri Marès of embattled Montpellier were members, as was the prominent naturalist Henri Milne-Edwards, an early, rare French academic champion of Charles Darwin's theories. Clément Duvernois, the agriculture minister, announced a public competition with a prize of twenty thousand gold francs "to be awarded to the one who finds an effective and practical means of defeating the new disease of the vine known as the phylloxera." The commission would act as arbiter, and would declare the winner – and here was the clever bit – "should there be one." "The Minister has shown a deep anxiety for the interest of viticulture," the commissioners declared soothingly. "His appeal to men of science will be heard and it is to be hoped that we will soon be in possession of a complete history of the disease and an efficacious remedy that will restore confidence to our *vignerons*."

It would not be that simple.

Part Two

ANGER

13

On 8 May 1870 a popular plebiscite in France overwhelmingly endorsed moves towards a "liberal empire." A new foreign minister was appointed, the Duc de Gramont, liberal enough at home, but a hard-liner and fatally clumsy when it came to affairs abroad. When opportunistic attempts by Berlin to install a German prince on the throne of Spain mysteriously surfaced, both the Duke and the political classes of France were gripped by a hot flush of patriotic outrage. On 16 July the Second Empire declared war on Prussia. After three weeks the French army was still not fully mobilized.

The armies of Prussia and her German allies were faster getting to the railheads. France's eastern fortresses were rapidly invested. After bloody battles on the frontier, almost half a million spike-helmeted soldiers poured through the Lorraine gap – getting between Paris and the two main French armies in the north and east.

Professor Victor Signoret had more intimate concerns. In the garden-laboratory of his agreeable suburban villa the entomologist peered into his specimen tanks. Tiny yellow aphids disported within, descendant generations of the live phylloxera sent from Bordeaux by Laliman. The professor had cleverly managed to propagate several leaf-galls from them on a Chasselas, a vinifera. But in an experiment conducted on a hot late-August day, unseasonally plucked from their protective blisters, the leaf-living phylloxera proved totally uninterested in being transplanted onto the roots. They refused to suck, and perished.

The professor had good reason to urge on his experiments. Marshal Bazaine's army was fatally bottled up in the frontier fortress of Metz. Marshal MacMahon's "Army of the Rhine," now dramatically joined by the Emperor Napoleon III himself, was being pushed northwards towards the Belgian frontier. Two German armies were advancing unopposed on Paris. On 2 September, MacMahon's army was routed at Sedan. The emperor surrendered on the battlefield the next day. The Republic was proclaimed at the Hôtel de Ville. The new Government of National Defense declared that the fight would go on.

Professor Signoret had correspondence to attend to. A package had arrived from America. It contained crumbling vine leaves and a dust of dried-up insects gathered in the fields of distant Missouri. By microscopic examination the distinguished entomologist could see a remarkable likeness to those sent to him which had been spawned in the Gironde. Lichtenstein had been right. They really were the same.

By 15 September the armies of Prussia and its German allies were encamped around Paris's fortifications. They would stay there for over four months. Desultory artillery shells crashed into the city. The supply of fresh vines from the professor's autumnal garden, indeed of anything edible, was becoming problematic. With the falling temperature, the last leaf-borne generation of the besieged professor's phylloxera obligingly migrated to the roots. As the encircled citizens of the capital adapted to the insect plan of their swarming besiegers, the now-underground aphids hatched, sucked root-sap, moulted, advanced to transient adulthood, laid eggs and died. Winter came. In their glass tank a few sugar-engorged survivors fell into suspended animation. The city began to shiver and starve. Unlike the captive aphids, encircled Parisians could not conveniently hibernate.

* * *

Charles Riley lost contact with his French correspondents. "The blighting effects of war have not only entailed untold misery to millions in France but have paralyzed scientific investigation," he wrote that winter of 1870–71, "so that at last accounts, M. Lichtenstein was in Spain and M. Signoret shut up in Paris. I was however fortunate to have received from the latter gentleman posted a few days previous to the investment of Paris, a letter stating that on examination of specimens of our gall-lice which I had expressed to him, he was convinced of their identity with the European species."

In a postscript footnote Riley added: "Since the above was written I have heard from M. Signoret through M. Lichtenstein. Nothing daunted by the siege, the former carried on his studies of this little louse and wrote by balloon,* that though he himself was reduced to [eating] cats, dogs, and horseflesh, the phylloxera, which he had in boxes, kept well and in good health. No doubt our enthusiastic friend finds much solace in thus pursuing knowledge under such difficulties."

The Second Empire had fallen. The German Empire was proclaimed. While the martial dramas were being played out in the north of France, winter had abated the phylloxera's advance in the south. In the early spring of 1871 the hibernating aphids began awakening in the warming soil of the Midi and the Gironde.

There remained a tremendous mystery. It was a root-sucking insect that was devastating the vineyards of the lower Rhône valley. Although a virtually identical creature had been found on the leaves of American vines, an underground form had

---

* The first balloon break-out was on 23 September 1870, by an aeronaut named Jules Durouf who carried 103 kilograms of letters and military dispatches. By 28 January 1871, when Paris surrendered, the French had launched sixty-five balloons manned by sailors and inflated with explosive coal gas.

never been reported in the United States. Why? Because no one had ever thought to look for it. When imported vinifera vines perished on North American soil it had been put down to the climate. Should a dead vine be exhumed, no underground attackers were found. The aphids had moved on.

In early December 1870 Charles Riley had gone looking. He had to travel only as far as Isidor Bush's fine establishment at Bushberg in Jefferson County, twenty-five miles east of St. Louis. In 1867 the nurseryman had conveniently imported from his native Austria-Hungary a number of rooted vinifera varieties, "not in hope of cultivating them with success," he said, "but with a view to discovering, by careful observation, why European vines failed to grow in North America and perhaps find a redress." His Tokays and Rieslings had "begun in a splendid manner," but in the summer of 1869 they became "yellowy and sick." By the following year the Danubian transplants were clearly dying. The entomologist arrived to see for himself. "With our full authority Professor Riley visited our vineyard to dig up both healthy vines and sick vines to examine their roots," Isidor Bush wrote.

Not all the guilty aphids had yet slunk away. The roots of the blighted immigrants were covered in the subterranean yellow plague. Sucking aphids were found when the roots of nearby American vines were examined, but in much smaller numbers – above ground meanwhile the vines stayed in exemplary health. "We thus see that no vine, whether native or foreign, is exempt from the attacks of the root-louse," Riley recorded. "On our native vines however when conditions are normal, the disease seems to remain in [a] mild state and it is only with foreign kinds and with a few of the natives . . . that it takes on the more acute form."

Isidor Bush thought it a historic moment. A two-hundred-year-old mystery had been solved. Not only did the discovery prove that the insects on both sides of the Atlantic were identical, but his visitor had uncovered "the principal reason why the culture of European vines has been an absolute failure in our country."*

Charles Riley arrived in France in July 1871 to see the insect cataclysm for himself. He went first to Paris for an entomological summit with Professor Signoret. A quarter of the French capital lay in scorched ruins.† Twenty thousand of its inhabitants had died in the bloody suppression of the Commune two months before. Mass trials of the insurrectionists were about to start. German soldiers still garrisoned the Paris forts. Some things had happily survived defeat, siege and revolution. The professor proudly showed his American visitor the "yet living progeny of some lice which he had placed in a tightly corked tube the year before – and that he managed to keep alive all through the siege of Paris."

Professor Signoret was prepared to be accommodating – yes, he might admit, the insect plaguing the Rhône valley was the same as the one his visitor had found on vine leaves in Missouri; but he clung to his first principle, that the phylloxera was sim-

---

* There were those in America who thought the opposite. William Saunders, the Washington superintendent of gardens, could write in the U.S. Department of Agriculture report as late as 1876: "the insect is falsely accused . . . They have long been observed on the roots of grapes; but it is only when the plants are otherwise diseased and their normal vitality impaired that the insects prevail to a fatal extent. The true cause is 'atmospheric influences' as is clearly shown by successful culture of vines in glass structures."

† The visit conducted so soon after the end of the siege obviously made a big impression on Riley, who returned several times to the theme in his writings. For example, in 1875 a huge swarm of Rocky Mountain locusts "ate up every green thing" in the Mississippi valley, and there were reports in Missouri newspapers of farming families dying of starvation. "Who can doubt that the French during the late investment of Paris would have looked upon a swarm of these locusts as a manna-like blessing from heaven and would have much preferred them to stewed rat?" the entomologist wrote. He proposed that the insects could be turned into a nutritious paste and stored for consumption in future times of famine, and published several recipes for locust – fried, roasted and stewed – cooked by his own hand in a St. Louis restaurant. They had a not unpleasant "nutty" flavor, said diners. Later that year he took his locusts to France for consumption by members of the Société Entomologique, who pronounced them quite edible fried in their own oil and served with a little salt.

ply a consequence of something else. It was "meteorological perturbation." Riley was baffled. "Aphidae, I repeat, must always be the cause rather than the effect of disease," he wrote sternly. He thought his host was "too much absorbed in closet studies to make the proper field observations."

Planchon and Lichtenstein were clearly happier to get out into the field. Riley ventured south to the hot plains of the Midi to go through the now well-trodden ritual of inspecting blighted vines. He saw the root-living form of the insect in all its horrible abundance, and noted winged phylloxera "caught in spiders' webs." Were they males or females? The evidence was contradictory. According to Planchon:

> In July 1871 M. Lichtenstein and myself at virtually the same moment were able to distinguish forms of the winged insects. One had a shorter abdomen . . . generally devoid of eggs . . . the other had a slightly longer abdomen containing two or three large eggs. The absence of eggs in the first case came to indicate the male sex, and the presence of eggs in the second the female . . .
>
> Our conjectures were shared by M. Riley, who having come to Europe this same year . . . would carry back some specimens of the two forms and himself find them in America the following year.
>
> However great doubts rose within us, especially about this so-called male. The absence of copulatory organs in both forms troubled us . . .

The aphid's sex-life would prove highly vexatious. The French would argue about it for years.

That the plague had come from America, Planchon himself now had little doubt. At long last he was apprised of Professor Westwood's observations (a photographic copy of the *Gardener's Chronicle* article of eighteen months earlier had been sent to him), which he clearly found fascinating. Was it not a "very remarkable coincidence," he wrote in July 1871, that first the *oïdium*, then the phylloxera had turned up in English graperies before making their "great ravages in the vineyards of France"? It was possible that the *oïdium* invasion had come by this route in the first place, he believed, as some had suggested but never proved at the time. Now the aphid, America's second fatal gift, menaced "all the vineyards of the world . . . demonstrating as so often happens with an imported epidemic, they are much more dangerous in countries to which they are strangers than in their native land." He advised growers in Burgundy and Champagne to beware of imported vines, "not just from the United States, but from England, Ireland and even Scotland."

There was another mystery. The wingless root-suckers were clearly the same insect as the specimens of leaf-gall inhabitants that Riley had brought with him from Missouri. But why were reports of galls so extremely rare on the vines of south-east France? On this field trip they had not observed a single one. Riley carefully preserved his Midi-gathered specimens in phials of acetic acid and blobs of glycerine to bear back to his Missouri laboratory for further examination.

"Here we have an insect, the life-history of which is as interesting to the entomologist as its devastations are alarming to the vine-grower," Riley wrote in November 1871. He clearly found the life and times of *Phylloxera vastatrix* absolutely fascinating – especially how it switched from an above-ground to a subterranean existence.

"By August . . . after the vine has finished its growth and the young lice finding no more succulent leaves, they begin to wander and to seek the roots so that by the end of September the galls are deserted," Riley noted. "Upon the roots the lice attach themselves singly or in little groups and cause by their punctures little swellings or knots which eventually become rotten. Strange enough, these lice not only change their residence as winter approaches . . . just like the Moor who having passed the summer on his roof, gets into his house, but Proteus-like they change their appearance in shedding their skins and all become tuberculed."

Riley suspected the insects could survive the winter in this "tuberculed" state, but what happened with the returning warmth of spring had yet to be resolved. Did they produce winged males and females which would "rise in the air to mate and thus produce the wingless mothers found in leaf-galls surrounded by their eggs"? Or did they lay "eggs on the roots, and the young hatching from these eggs crawl up onto the leaves and found those gall-producing colonies"?

Both of these theories would turn out to be wrong. Whatever the answer, it was plain to Riley that the "insect can be transported from one vineyard to the other on roots. Doubtless it was by some such mode as this that the insect was introduced into France from this country." On that presumption he was absolutely right. Planchon had come to the same conclusion. The trouble was, very few in France yet thought the same.

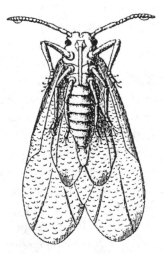

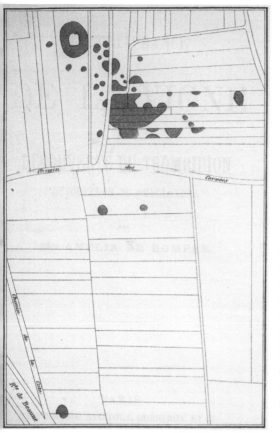

Left A map of infected vineyards around Dijon shows the "oil-spot" pattern of the malady's spread which investigators at first found so baffling.

Below When vines in the lower Rhône valley growing along river-banks that flooded in winter were observed to survive the malady, it was realized that inundating vine-yards would drown the under-ground insect. The submersion technique worked on the plains, but not on hillsides. Where practi-cal and affordable, vine-growers built embankments and used steam pumps to stifle their enemy.

When the phylloxera first struck, carbon bisulphide was already known to be an effective insecticide. The problem was how to get the explosive chemical, which vaporized at room temperature, deep enough into the ground. Giant cast-iron syringes were devised, and the P-L-M railway company recruited a small army of technicians known as *piqueurs* to wield them. Growers were encouraged to join syndicates and were offered subsidies to use the chemical, while in some places "administrative treatments" were made compulsory. Skeptical vignerons, however, reacted with either indifference or outright resistance.

Right Research into effective insecticides was a high priority. Hundreds of bizarre proposals were advanced in bids to win the state prize of 300,000 gold francs for an effective and practical remedy. Large-scale field trials were conducted, while this "death-chamber" was promoted as an ideal means to find what might kill the insect.

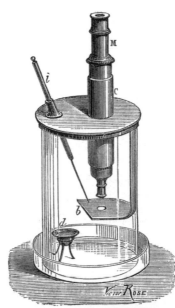

DAVIS' COMPOUND INSECT DESTROYER

Desperate vine-growers grasped at whatever opportunist manufacturers might proclaim as sovereign remedies for the plague. Mr. Davis of Boston came to France in 1875 to promote the merits of his "compound insect destroyer." Just what was in it remains mysterious.

The chemical potassium carbonate, more expensive to manufacture than carbon bisulphide, was found to be an effective insecticide after research work largely sponsored by the brandy-makers of Cognac. It needed large quantities of water to get it into the ground; only grand proprietors could afford the necessary steam pumps and pipework.

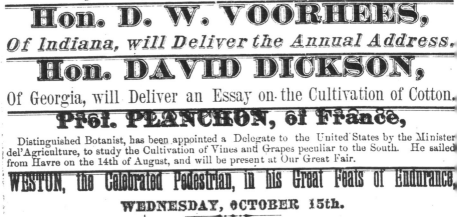

Special Trains for Passengers will run to the Grounds from the City every fifteen minutes. Fare only ten cents.

Hon. D. W. VOORHEES,
Of Indiana, will Deliver the Annual Address.

Hon. DAVID DICKSON,
Of Georgia, will Deliver an Essay on the Cultivation of Cotton.

Prof. PLANCHON, of France,

Distinguished Botanist, has been appointed a Delegate to the United States by the Minister del'Agriculture, to study the Cultivation of Vines and Grapes peculiar to the South. He sailed from Havre on the 14th of August, and will be present at Our Great Fair.

WESTON, the Celebrated Pedestrian, in his Great Feats of Endurance,

WEDNESDAY, OCTOBER 15th.

In 1873 Professor Planchon was sent on a state-sponsored mission to the United States to further research the life-cycle of the insect and the potential of American vines to resist its attacks. At the North Carolina State Fair at Raleigh that October an appearance by the "distinguished botanist" was high on the bill.

14

§

In August 1871 the Central Commission delivered their first report, which had been delayed by war, to the new republican government in Paris. It was printed in a popular edition with illustrations for the edification of vine-growers, and run in the *Journal officiel* and other newspapers. Their findings provided an admirable statement of all that was known of the aphid's life-history. The report made no guesses as to where it might have come from. As to combating it, the learned commissioners could only offer pious hopes:

> *The most characteristic external feature of the new disease, the origin of which is unknown, is the existence in all recently affected spots of a center of attack which extends itself without intermission.*
>
> *The vines contiguous to the first seat of infection, shed their leaves and grow yellower and yellower until they become completely dried up ... Experience has also taught us that the new vine disease progresses by irregular bounds, often appearing suddenly at great distances from the center of infection.*
>
> *These serious disorders are due to a species of* aphis *which has been named* Phylloxera vastatrix. *This insect which is almost invisible to the naked eye establishes itself upon the roots of the vine and perforates them with its proboscis in order to feed upon their juices.*
>
> *It must be remarked that the phylloxera never remains*

upon the roots which are beginning to decompose. When one part grows putrid they immediately remove to another.

The phylloxera has two different phases of life. It nearly always remains below the earth, though now and then a few individuals are liberated to the enjoyment of an existence in the open air . . . It would be very useful if we could only ascertain precisely at what time of the year the transformation of the winged insect takes place and how long it remains alive . . .

The phylloxera is not viviparous [live-young bearing] but throughout the whole season and in both forms it lays only eggs. Individuals hitherto observed have all been female. The male phylloxera has not been found although long diligently sought after.

They pass the winter on the roots in the wingless state and never in the egg condition . . . they remain in a state of perfect torpor; but as soon as the warmth begins to make itself felt, all those individuals which have survived the cold and the damp of the winter begin to wake to renewed life. They feed with great avidity and immediately begin to lay eggs.

During the warm season a few individuals . . . throw off their covering and appear as perfect insects provided with wings. It is in all likelihood after it has taken this form that the phylloxera is borne up and carried away by the wind, though it cannot be affirmed that even the wingless ones, in certain conditions, are incapable of being thus distributed.

The winged form was "very rare," said the report.

The number of these which have hitherto been observed bear no proportion to the myriads of wingless insects that can be seen in every part of the affected roots.

Can this be natural? . . . or is it an error of observation? All the winged phylloxeras which have been seen were females, who lay eggs and thus give birth to the wingless lice.

The report mentioned "a fact of very great importance," the discovery of leaf-galls in the Rhône valley, "and still more so in the vicinity of Bordeaux." These were evidently due to "the agency of the winged phylloxera."

Let us hope that diligently prosecuted biological studies will soon enlighten us upon these strange and momentous questions. It will perhaps be possible to destroy this insect which is so very troublesome to get at during the time it lives underground, if we could only obtain a favorable chance of catching it on its excursions in the open air.

The Commission counsels vignerons and municipal bodies to follow the example given in the Hérault and the Gironde where the diseased vines have been rooted up and the soil disinfected by the burning of the surface.

Applied in time and systematically conducted under intelligent supervision this may impede the progress of the evil and even repel it.

It would have done no such thing. And it was by now far too late anyway. Planchon pointed out the inherent unworkability of a mass eradication strategy when the "evil stayed invisible underground" on apparently healthy vines. He and Lichtenstein had argued almost from the beginning that uprooting and burning only made sense if applied to the very first spots of infection.

"Some savants have claimed it can be eliminated just as sick cattle are slaughtered by law to fight bovine plague," Planchon wrote, "but the law cannot be applied to a subterranean vermin. The problem is how to apply the judgement . . ."

The picture the report painted meanwhile of self-sacrificing *vignerons* setting the torch to their vineyards was just not true. The wagons mournfully trundling the autumnal Midi laden with dead vines were more often heading for the market square to sell them as firewood. This lowly trade was beyond control.

Victor Lefranc, the minister of agriculture, agonized. At this mournful stage of French history, "invaded departments" referred to those still occupied by the Germans – not those under the aphid's thrall. A scorched-earth policy imposed by Paris was politically impossible. In September the minister circularized prefects with the Central Commission's report, feebly exhorting them to "bring it to the attention of the principal wine-growers of your department." "Although these prescriptions have no compulsion," the minister wrote, "I urge you to press on the municipal authorities of your department – which are or which will soon be invaded by the plague – the necessity of applying radical measures without hesitation . . ." This was hardly "They shall not pass."

The prefect of the Hérault had on his own authority made uprooting and burning compulsory in 1872, and banned the import of "foreign vines" – but his decree was widely ignored on both counts. The mayor of the commune of Fontes for example reported that "a few parcels of vines in the territory . . . are infected . . . but the proprietors of the said parcels, rather than burning them on the spot . . . transported them hither and thither thus propagating the disease."

There was an attempt to raise subscriptions from wealthier proprietors to pay compensation to growers who agreed to dig up and burn their vines. The response was feeble. Many refused the offer of money anyway. In 1871 the *Messager agricole du Midi* reported the attitude of one grower urged by representatives of the Hérault agriculture society to dig up and burn his dying vines: "Although an indemnity would have been paid to him . . . the proprietor defended his plants with a quite fatherly resistance," reported their emissary. "He did not understand at all the importance of the operation we wished to attempt . . ."

* * *

Uprooting a mature vine is not easy. Windlasses, ropes and pulleys might be necessary. Another technique was to use a substantial tree branch with a cleft fork at one end. Pulled by two men over an improvised log fulcrum, it worked like a giant crowbar. It was known descriptively as *le chèvre* – "the goat." Later a wheeled derrick was devised with a grab hook. It looked like a mobile gallows.

Some altruistic Hérault proprietors like M. Pommier at Lunel-Viel did admit the labor gangs armed with their horrible forked branches and cans of petroleum, "in the interests of neighboring vineyards." But the aphid obeyed no proprietorial boundary. As Planchon realized, to be effective all the "healthy" vines around a center of infection had to be uprooted and burned – impossible to enforce, given peasant stubbornness. Gaston Bazille yearned to uproot the "doubting Thomases" and march them just a few kilometers to see for themselves the devastation remorselessly bearing down on their communes.

The perplexing geometry of the aphid's advance inspired fitful hopes of salvation. A close-textured analysis of terrain, insect and human frailty reveals why. The first appearance in the Hérault was in 1867, in the eastern *garrigue*, around the commune Saint-Bauzille-de-Montmel. Two years later the initial outbreak had expanded aggressively outwards, there were breakaway oil-spots on the coast, and a large satellite outbreak at Villeveyrac thirty kilometers to the west.

By 1870 the malady had descended the eastern heights and was ravaging the vineyards of the plain from Lunel to Maugio. The following year its progress was marginally less rapid (the summer was cool), but in spring 1872 it woke up with renewed voracity, advancing rapidly up the coastal plain and along the course of the Hérault river-valley. Several uninfected islands were left on hilltops. It seemed perhaps that hill ranges might act as some defensive barrier. This was another mythical life-raft to cling to: some infected communes lost all their vines within two years; in other places the localized mortality was much slower in its coming.*

A twentieth-century analysis explained it this way: "The close proximity of the vineyards in the plain enabled individual winged phylloxera to spread easily and quickly over the entire area, the deep and close-packed soils in which the vines were planted denied such ease of movement to the much more numerous phylloxera of the roots." Yes, the airborne plague was everywhere – but the speed and power of the insect's destructiveness underground depended on highly localized soil conditions.

The stop-start rhythm of the plague's diffusion explained in part the human response. Peasant intransigence was understandable given the social texture of rural France, where the aphid might travel further in its short lifetime than a *vigneron* would venture from his commune in his. But there was more – an absolute denial that the plague would ever arrive, even when dying vines were just over the next hill. "The population of this rich department does not wish to believe that the scourge which menaces their vineyards could lead to the almost complete destruction of their vines," wrote Frédéric Cazalis in the *Messager agricole du Midi* in summer 1871. "I am assured that in the *arrondissement* of Béziers [in the west of the department] there are some people for whom the phylloxera is almost a mythical beast," wrote the *Messager*'s correspondent two years later. "One can hardly credit such disbelief scarcely a few kilometers from the ravaged areas." The vineyards of Béziers would all succumb.

Planchon was baffled by the strange pattern of advance. Without a full description of the aphid's life-cycle (the elusive

---

* The respite did not last long. By 1878 the entire eastern half of the Hérault was infected, and the hilltop outposts had succumbed.

male had yet to be found) it could be assumed that the winged aphids in their two microscopically different forms founded new colonies by directly laying eggs either on leaves or on vine-footings. But something was missing. So few aeronauts had been found. And their wings seemed too fragile to sustain flight for any distance. In his Montpellier laboratory in the summer of 1872, Planchon had observed captive females "begin to vibrate their wings excitedly, launch themselves on an oblique trajectory for a few tens of centimeters . . . then fall onto the table – like a kite suddenly deprived of wind." Their abdomens bore eggs, two or three each. What did they hatch into? "In captivity we have seen her lay these eggs on the sides of her glass prison," Planchon wrote, "then she dies after two or three days of existence." The eggs were sterile.

"Placed on a vine branch under a bell-jar she is dead before having laid any eggs or without having pricked the leaves in any discernible manner," he noted. "We have no idea of where these eggs are laid [in a natural environment], what kind of insect emerges from them, and what relationship these new-born have with the leaf-galls on certain vines or with the roots on others . . . Answering these questions is the most important *desideratum* in unravelling the life story of the phylloxera."

Professor Planchon proposed to do so, if necessary by joining Riley in America.

The mischievous nephews of Louis Faucon had seen top-hatted phylloxera out on a summer's day strolling along the ground with walking sticks. No one had believed them. In August 1872 the boys saw the yellow strollers again, but this time members of the Central Commission were there to confirm the phenomenon. Remarkably, a number of winged individuals were observed to join the aphids' scuttling progress across the clay soil at Graveson.

The discovery was hailed as a great breakthrough – the

plague, it seemed, might somehow be fought on the surface. But Planchon and Louis Vialla quickly proved by observation and simple experiment that the aphids were able to sniff out food and move from vine-root to vine-root through underground cracks in the soil. Dry clay offered especially navigable thoroughfares.

That would certainly solve the "oil-spot" puzzle. But if the winged phylloxera that Planchon had observed were such poor fliers, how did the infection spread in kilometer bounds? He was moving to a new supposition. "It is possible that the winged form is not the only way that it can be transported by the wind," he proposed in 1873. The wingless crawlers on their curious marches across the surface of the ground "could be picked up and blown long distances by violent winds," he wrote, "but this is pure hypothesis." The botanist was horribly correct.

15

The outbreak of the new vine malady was of more than entomological interest to those who consumed the products shipped
from the quaysides of Bordeaux. In the summer of 1872 the
British Foreign Office began to take an abiding interest in the
aphid. It was not just that vine culture in Britain's colonies
might one day be afflicted: supply of those staples of upper-
middle-class English life, claret and port (the first infection in
the Douro valley had been reported a year earlier), might not
just be interrupted but entirely extinguished. The *Wine Trade
Review*'s Bordeaux correspondent reported that August: "The
weather is magnificent and there is every reason to think the
quality of the vintage will be excellent . . . it is to be feared
however that the quantity of the crop will be lamentably deficient, due to the ravages of the phylloxera."

The threat was taken very seriously in London. The Phylloxera Commission's first report was translated in full for the attention of Earl Granville, the foreign secretary. Dr. Joseph
Hooker, director of the Royal Botanic Gardens at Kew, was
brought in to give his expert opinion.

Sir Charles Murray, HM minister at Lisbon, became alarmed
when the vine in the embassy garden began to "wear a sickly
appearance," as he informed Earl Granville in July 1872. "The
leaves began to wither and assume first a yellow and then a
reddish appearance." The ambassador's gardener was perplexed; the vine was exhumed, but when examined the roots
"appeared perfectly healthy." Sir Charles had heard that a

"French vine-proprietor had tried with some success the expedient of digging a hole round the vine which he then fills with chimney soot . . . it is much to be regretted that the vine-growers of France and Portugal are not somewhat nearer to London, where it could be cheaply and abundantly supplied," he informed the foreign secretary unhelpfully.

Consul Oswald Crawfurd in Oporto was asked to investigate further. He ventured inland up the Douro valley into the region where the uniquely precious monoculture of grapes for port-wine making was conducted on steep hillsides on which a scrap of flinty soil was so valuable that precipitous terraces had been chiselled out by hand or blasted by explosive. The consul seemed skeptical: "French oenologues had not a little over-stated the extent of these ravages," he reported – while in the Douro itself the publication of a pamphlet by a certain Senhor Oliveira based on excitable reports from Montpellier had "encouraged undue alarm."* Seeking moisture in such poor, arid soil, Portuguese vines were too deep-rooted to be vulnerable, the consul had been assured, while those vine-growers he had himself spoken to had seen no unusual insects. If there were problems elsewhere it was due to "bad air."

The vine-growers of the Bordelais seemed even more blinkered. According to Duclaux's Paris-drafted invasion map, the first infected spots around Léo Laliman's domain near Floirac had already splashed eastwards in the direction of the prevailing wind. By the summer of 1872 the aphids had broken out of the Entre-deux-Mers, crossed the river Dordogne and were advancing on Libourne. But when asked to investigate further, Thomas Carew Hunt, HM consul at Bordeaux, could

---

* It was however the beginning of a national catastrophe as profound as that which would engulf France. A hundred and thirty years later, the higher flanks of the Douro valley remain corrugated by empty stony terraces, left as they were in the 1870s, denuded by the phylloxera. They are pointed out to visitors as *morotórios* – "places of the dead."

inform Lord Lyons, HM minister at Paris: "The introduction here of an insect of such enormous multiplying functions would have been a constant subject of reference by the local newspapers." He could find only one mention – a recent soothing article in *La Gironde* stating that when roots had been dug up on dead and dying vines at Floirac scarcely any phylloxera were found. Surely some agency other than this "microscopic *hemiptera*" was responsible, the newspaper had commented.

The consul inquired of M. Corti, president of the Bordeaux Chamber of Commerce, what he might know. The spokesman for the *négociants* of the great wine entrepôt seemed supremely unworried. The site of the outbreak at Floirac "had not notably expanded," Corti assured Hunt. In spite of the offer of the twenty-thousand-franc prize, "no one knew the causes which have produced this malady . . . and as for an efficacious cure, there is none." The advice of the Gironde Société d'Agriculture apparently was to dig up "dead and sick vines and plant new ones . . . several proprietors around Floirac have followed this advice, and today their new vines are beautiful." So that was that. Corti meanwhile felt compelled to express his doubts whether the phylloxera was really the cause of the malady, or simply its consequence. Indeed, whatever had afflicted Floirac, "was it really the same as the malady so generalized, so redoubtable in the Languedoc?"

M. Dupont, secretary general of the Gironde Société d'Agriculture, proved supremely untroubled. Those who believed "unproven rumors" about the presence of the *puceron* were making "a profound and grave error," he insisted. Recent disasters with certain vines were caused by "atmospheric vicissitudes accentuated by accidents of geology." The whole "question of the phylloxera" was some sort of hoax, so Dupont implied, invented by "publicists" for their own self-aggrandizement . . .

* * *

This was the reaction in a department where the aphid was already on the march. There were many more Frenchmen for whom the troublesome aphids, if they existed at all, might as well have been some tribal insurgents in a distant colony. When the first cries of anguish had come from the Midi, the *Journal de Villefranche* expressed the view of many in Burgundy's southern rampart of granite hills, the proud Beaujolais. The plague was "a just punishment" brought down on their greedy southern neighbors for their "bloated production," the newspaper editorialized. Local sages proclaimed that the upland climate of their fortunate region would save them should the insect, "spawn of the Midi and the hot plains," dare to outflank Lyon and venture northwards into their winter-frosted *collines*. It would dare – but not quite yet.

Planchon despaired. "The spirit of solidarity is totally lacking," he told delegates to the Lyon Congrès Viticole held in September 1872, when the aphid's *avant-garde* was just a few kilometers south of the city. "Yet we are all one. If your neighbor is struck today, you can surely say it will be me tomorrow. Help your neighbor, that is the only salvation. I believe it is my duty to address you thus – not only the people of the Lyonnais, but those of Burgundy, whose wines are one of our nation's glories. If Burgundy should be wiped out – along with Bordeaux – one could say that France itself had been overthrown." There was tumultuous applause.

16

Professor Planchon's call for solidarity may have gone down well with the endangered vine-growers of Lyon, but outside south-east France he and his little band of colleagues, Bazille, Lichtenstein, Vialla and Sahut, were derided as the beetle-hunters of Montpellier. In Paris, Professor Signoret clung to the phylloxera-as-effect theory. The Ministry of Agriculture preferred to listen to him. The metropolitan press stayed aloof. Mention of the aphid in the Assembly was scant.

The conservative government of Adolphe Thiers, suppressor of the Commune, had other concerns – steering delicately between republican and the majority monarchist deputies in the Assembly, remaking an army, raising by two monster bond issues the bruising five billion gold francs war indemnity demanded by the German Empire as the price of evacuating its armies. The as yet unphylloxerated vineyards of Alsace had been claimed by the victors.

The brave-sounding Phylloxera Commission meanwhile did little but meet fitfully to consider the increasingly bizarre proposals inspired by the imperial government's prize. The offer of twenty thousand francs for a practical remedy had survived war, defeat and revolution. French scientific and intellectual life continued. The Academy of Sciences published unbroken its biennial *comptes rendus* in which the latest twists in discovering the aphid's life-history were dryly reported.

Where the phylloxera reigned, however, there was an explosion of argument in the provincial press. Planchon propa-

103

gandized noisily in a series of pamphlets published in Montpellier, and in the columns of the *Messager du Midi*. Later the Parisian political journal *Revue des deux mondes* would make him its phylloxera star. The botanist's outpourings were a mix of empirical science, homespun philosophy and despair at human nature. "Facile arguments devoid of facts need only a little puff of rhetoric to seduce not only the ignoramuses, but many educated and sensible people," he wrote in 1874 as denial and fantasy crowded around the matter of the "terrible plague." He was right to be so cross. The battle against the aphid was being lost. The battle to change minds was just beginning.

There were plenty of new theories. There was the briefly fashionable concept of "degeneration," for example – that years of continued vegetative propagation from mother vines had excluded "male vigor" from the generations. This argument had been advanced before, during the potato blight and *oïdium* crises – and would be advanced again when scholarly debate was joined five years later on the very peculiar sex-life of the phylloxera.

Planchon thundered back: "How can one believe that simultaneously, at different points all over Europe both in greenhouses and outdoors, the sturdiest as well as the most delicate vines suddenly became 'degenerate'?" The invisibility of the attacker made people blind, he argued. "If a wolf eats a sheep or a caterpillar eats a cabbage, nobody is going to claim the sheep or the cabbage have suddenly become predisposed to being eaten." No more so had France's vines all at once decided they were ripe for being devoured by an insect. As for other supposed causes – "injudicious pruning," "exhausted soil," winter cold, summer heat – the botanist found counter-arguments to demolish each one in turn.

There were those who claimed that the phylloxera was an old foe in modern form. A certain Mr. Koressios argued in the Athens newspaper *L'Éclectique* that phylloxera was none other than the "*phtheir* or vine-louse, described in antiquity by Strabo, and that the modern Greeks still fight it by means familiar to the ancient geographer." And M. Nourrigat, president of the Comice Agricole of Lunel near Montpellier, insisted in the *Journal de Lunel* that the claimed new disease was the same as a malady that ravaged the Danube basin between 1730 and 1776. Then it had been called *Gabel* ("fork") because of a curious branching in the shoots of afflicted vines. Planchon robustly debunked these theories: "Can one imagine such an insect not showing itself for centuries and then suddenly breaking out with such devastating power?"

Léo Laliman of Bordeaux, stung by accusations that he was the instigator of the first fatal import, also unsurprisingly argued that the aphid had long been endemic in Europe. It was a pest of fruit trees, that had taken up residence on vines because of the excessive hunting of birds. To prove his innocence he invited an inquiry by a special commission of the Gironde Agriculture Society (with himself as a prominent member), which took an affidavit from Mr. J.P. Berckmann of Augusta, Georgia. The missive from Fruitland stated: "In our region the phylloxera is unknown; vintners do not dream of its presence, and the damage that it does is not worth speaking of."

Laliman added, unhelpfully perhaps for his own case, "that he had distributed American varieties through divers countries where the malady is unknown without introducing it." The commission acquitted him completely.

Surely it was obvious that the phylloxera had somehow come from America? Professor Planchon set out to prove it beyond doubt. Like an intelligence officer collating dusty reports of a terrorist sleeper-cell's awakening, Planchon worked out where the transatlantic visitors had first arrived. There was the notice by Professor Westwood on the Hammersmith and Cheshire outbreaks, and the statement (reported in the accounts of the

French Academy of Sciences in September 1872) of Malcolm Dunn of Powerscourt, Ireland. There was the evidence from the Gironde – where Léo Laliman had been importing American vines for years. There was recent news from Herr Rössler, director of the imperial government oenological research station at Klosterneuburg, near Vienna, which had been experimenting with American samples transmitted via a nurseryman in Hanover. The imports were prospering – the Austrian natives were dying in droves.

As for the lower reaches of the Rhône valley, where Planchon had first peered at the mysterious yellow clusters at the Château de Lagoy, there were several possible first points of entry. The botanist's suspicion fell on Tonelle, near Tarascon, site of the once-magnificent nursery of the brothers Audibert – "always full of exotic plants, many of them directly imported from the United States, this once beautiful establishment possessed twenty-seven varieties of American vines from 1838 onwards," as Planchon wrote. It was from Tarascon that the enterprising James Busby had obtained his rooted vines to be sent to New South Wales in their moss-packed, oil-paper parcels.

Now, thirty years later, the nursery had been obliterated. Planchon interviewed former customers. A certain M. Reynier of Avignon, an "intelligent and educated gardener," was clearly a good source of information. He was an old friend of the long-departed proprietors, and in the early 1860s had bought some American vines from them, Isabella and Catawba. In the spirit of amateur viticultural solidarity he had passed samples on to M. Clerc, the mayor of Roquemaure.

M. Borty of nearby Pujaut also had a tale to tell. In April 1874, in some sort of quest for absolution, he discreetly informed Paul Druysset, editor of the *Messager du Midi*, of his guilty secret. Twelve years earlier he had imported 154 American vines through the intervention of "a friend." The editor called in Planchon and two members of the Agricultural Society of Vaucluse to investigate. They spent a mournful if instructive day. "Monsieur Borty, a wine merchant, cultivated his own beautiful vineyard near Roquemaure which became the prey of the *oïdium*," Planchon reported to the Academy of Sciences. "Having heard that American vines escaped the attention of this cryptogam, M. Borty . . . acquired a number of American varieties around 1862 (the date is not certain). He made a little square plot of Clinton and 'Post-Oak' in the middle of his French vines. Today, perhaps twelve years later, in spite of having phylloxera on their roots, not only are they alive but luxuriant and full of vigor. Of the French vines planted in the same ground, a great number are dead or dying . . ." Planchon had no doubt that he had found the first outbreak. "One cannot but believe that it was these vines that were the vehicles that brought the phylloxera to our region of the Midi, just as it was those of M. Laliman that brought it to the Bordelais."

This was not so much to arraign the guilty as to prove the scientific point. Plant an American vine, and sooner or later every vinifera around it would die. But why had it not happened before? Exotic vines had been imported from the Americas for centuries. Planchon burrowed into the archives, noting an 1823 article in the magisterial *Cours complet d'agriculture* by the naturalist Louis-Auguste-Guillaume Bosc, keeper of the royal nurseries in Paris of King Louis XVIII.

Bosc had mentioned American vines cultivated in them, including a Vitis cordifolia, one of the species discovered on his journey of botanical discovery by André Michaux. "The vines of this variety had died as a result of some destructive force that had swept over them," the royal gardener had written mysteriously. "This rather obscure sentence might imply that conditions in these nurseries were bad generally for the culture of plants," Planchon commented, ". . . but the death of the Vitis cordifolia could also have been caused by the phylloxera imported perhaps from America." The outbreak fifty years previously was contained, Planchon proposed, because there were no surrounding vines on which it could spread. He admitted all this was guesswork, confused by the fact that "varieties derived from Vitis cordifolia are those that resist the insect." "Numerous vine shoots have presumably come from America without being infected," he thought, "but perhaps a rooted cutting was the fatal import. It would be unjust to present this as some sort of crime because the authors of such actions were completely unconscious of its consequences."

Everyone was innocent. If there was a guilty party (although Planchon himself did not make this assumption), it was the engineers of mid-nineteenth-century Europe and America, who by the early 1860s had made steam-powered cargo ships an efficient commonplace on the trade routes of the north Atlantic. A three-week crossing under sail was now a matter – by the fastest steamship – of ten to twelve days. According to one prominent theory, "early vine introductions are likely to have been cuttings, rooted cuttings and seeds. If they had been infected with aphids, they would have died by the time the long sea voyage was completed. But steamships carried the plants far more quickly and the railway reduced the time of the inland voyage." The theory is attractive. To test it by experiment a century and a half later would be illegal.*

Cocooned in nurserymen's oil-paper parcels, root-living phylloxera nymphs might indeed have found enough nutrition to sustain life and growth long enough to make the journey from Missouri to the Midi. And because they were travelling on American roots their numbers would have been very small,

---

* Under Statutory Instrument No. 1758, the Plant Health (Great Britain) Order 1987 Part II, Article (1), "plants which may not be landed if carrying or infected with the plant pests specified" include: "*Daktulosphaira vitifoliae* (Fitch) (syn. *Viteus vitifolii* (Fitch)) – Grape phylloxera, Plants of *Vitis L.* other than fruit and seeds." It is also a notifiable pest. According to HM government instructions: "If you see or suspect the presence of phylloxera you must immediately inform your local DEFRA Plant Health and Seeds Inspector."

making them invisible as they went hungrily into the soil of the
Rhône valley to sniff out the delights of defenseless vinifera.
Unfortunately, there are problems with the theory. Hibernating larvae or overwintering eggs would not need to eat anything
at all. On certain American species of vine, the time between
egg laying and the emergence of phylloxera can be over thirty-
five days. Early importations into France in the 1820s and
1830s may have just mysteriously affected a few plants and
then died out. It was the transportation of rooted samples that
was the vector for the parasite; that and the sheer scale of imports from the mid-1850s onwards, ordered by vine-growers
as a prophylactic against the oïdium.*

By 1872, ironically perhaps, a new tide of transatlantic vines
was on its way to south-east France by steamship. The message
of Messrs. Planchon, Laliman, Bazille and others had taken
root in the Midi at least: American vines were somehow immune to phylloxera. No other means of defense as yet advanced
had proved effective. The esteemed Mr. Charles Riley, *entomologiste en titre* of St. Louis, had published his observations,
gathered from Missouri farmers, on which American species
showed resistance and which did not. With the war barely over,
in February 1871, the *Messager agricole du Midi* published his
view that the "frost grape" (Vitis cordifolia) was vulnerable.
"No variety of it should be cultivated and encouraged where
those of the fox grape (V. labrusca) or the summer grape (V.
aestivalis) are known to be as good," he wrote. (The Atlantic
mails had been re-established, and Jules Lichtenstein did the

* For example, the *California Farmer* of 31 July 1857 noted as the oïdium crisis
deepened: "Mrs. S.J. Kellogg of Cincinnati, who resided for many years in France,
has lately received an order from Bordeaux for cuttings *and roots* [author's italics]
of our American varieties, and Col [*sic*] Marshall P. Wilder of Boston has been
commissioned by the government of Belgium to send over all our best grape
vines . . ."

translation from page proofs of a December 1870 article in the *American Entomologist.*) Clintons would fail; Concords would prosper.

The sagacity of Mr. Riley was now further proclaimed in the catalogue of Messrs. Bush & Sons with its colored illustrations of luxuriant grapes and novel instructions on how to order vines by mail. Copies dispatched to Montpellier from St. Louis were like gold dust. Those who had seen it were enraptured. The company was even generously offering to send trial samples to new customers *gratis*.*

Riley had admitted elsewhere that his observations might be flawed, as "botanists and experienced grape-growers [in America] could not agree on the number of indigenous species of the grape-vine and the true character of the cultivated varieties." Why, some learned gentleman insisted there could be only one species of the grape-vine, if the Darwinian precept of infertility of interspecific hybrids was taken as the test.

The embattled proprietors of the Midi had no time for such sophistries; they wanted vines. It was demanded by some that warships of the French navy should be dispatched to New York to get them. In 1872 Victor Lefranc, the minister of agriculture and commerce, was petitioned to overturn the prefecturial ban and allow the mass import of American stocks. He willingly agreed, and the consul in New York City was instructed to assist in the huge operation of shipping them by railroad and steamer to Le Havre. One big proprietor, Jean-Henri Fabre,

---

* The Bushberg catalogue would later be translated by Gaston Bazille, with an adulatory introduction by Planchon. It was unashamedly evolutionist in both French and English versions: "We see in the general resistibility of our purely native American vines against the phylloxera, a remarkable verification of the law which Darwin has so ably established and aphoristically expressed as 'THE SURVIVAL OF THE FITTEST.'"

† Jules Lichtenstein claimed later that he had gone to Paris "in the winter of 1870–71" to personally urge the minister to expedite the import of American vines via the French government's consular agents. How the scientist got through the Prussian siege lines he did not explain.

magistrate and former imperial senator, ordered a huge quantity of Clinton and Scuppernong vines from Bushberg to replant his by now totally devastated vineyard at Saint-Clément near Montpellier. There were no care instructions sent with the American vines, no guarantees they were going to work. It was up to Fabre and those like him to discover if they would take to French soil in all its subtle variations, let alone make a wine that was drinkable. The science was incomplete.

The result was anarchy – more imported vines and more imported aphids. It was the beginning of a new disaster.

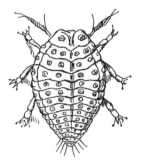

17

❧

In the early days of the aphid's march, M. Duvernois, the imperial minister of agriculture, had expressed his hope that his "appeal to men of science will be heard . . . and we will soon be in possession of an efficacious remedy." His republican successors in office (there were six in the space of three years) had so far gone unrewarded.

Professor Planchon and Camille Saintpierre had gotten nowhere with their 1869 experiments at Rochet, made in the first flush of panic to find something that would kill the insect deep underground without killing the plant. Uprooting and burning had failed. Nor had holy relics brought forth a miracle. There were those who still looked for divine intervention. Setting vials of holy water from the shrine at Lourdes among the invalid vines was one proposal.

Suggestions from "science" were no better. Since first being offered in spring 1870 the government prize had brought forth, in Planchon's words, "an avalanche of grotesque *élucubrations* which would be laughable if they were not a humiliating symptom of the failings of our national education." Proposals for remedies were supposed to be sent to the Ministry of Agriculture with an assurance that they had been "tested and found successful." They were then to be forwarded to the scientists on the Phylloxera Commission for some sort of official trial. That was the plan. They had arrived in the hundreds from all over France and beyond. Some were inky scribblings, others obsessively worked-out results of bizarre experiments. The more serious merited mention in the biennial reports of the Academy of Sciences: "Communication sur le phylloxéra de Monsieur x x x," with a follow-up reference in the *Journal d'agriculture pratique.* Most were absurd. Roots should be drenched with white wine or an emollient herb tea made from flowers of mallow. Living toads should be buried in blighted vineyards to draw the poison from the soil. "Strong-smelling" plants such as hemp, lupin, juniper, wormwood or box should be planted as repellent cordons. American vines should be planted amidst the European, as the insect would prefer them (a fatal suggestion). Crops such as red corn (known to be a favorite of other aphids) should be planted amid vines to draw the parasite onto them. "Nonsense," said Planchon, they had got entirely the wrong insect. Such ideas were the outpourings of *"fantaisistes."*

Proposals for what today would be called "biological control" provided an early avenue of hope. Jules Lichtenstein had been advocating such a thing almost from the very beginning. Was there an animal or plant in France or America that might prove the natural enemy of the unnaturally transplanted aphid? Planchon was in the audience when a whole afternoon at the Lyon viticultural congress of September 1872 was given over to a presentation on the advantages of *Madia sativa,* a curiously viscous plant native to Chile, commonly known as "tarweed." Planchon must have despaired. M. Maudit of Chartres, champion of the South American plant, was borne on wings of rhetoric:

> *Gentlemen. These are terrible times. Pity the generation –*
> *we ourselves – who must live through them. It was not*
> *enough for us to have faced the disasters of a war which*
> *began with defeat, but which will surely end in triumph.*
> *It was not enough that revolution should stain our cities*
> *with blood and a dreadful murrain decimate our cattle . . .*

> *A hideous plague previously unknown had yet to descend on us* – Phylloxera vastatrix – *an invisible enemy that is today in incalculable numbers remorselessly attacking our most famous vineyards, bringing despair and ruin where once there was happiness and prosperity.*

The delegate portentously expounded on what was known of the aphid's life-cycle, presuming that the winged phylloxera (also the wingless form if blown by strong winds) was the harbinger of long-range doom. He had noticed meanwhile that abandoned vineyards, now invaded by weeds, were "free of the plague." There was another observation: the insect reigned in the south, where it was hot, and not in the north. If the Midi's febrile soil could somehow be cooled down, the insects would surely depart.

Tarweed was the magical answer. Affordable by the humblest *vigneron*, it was far more effective than the Venus flytraps or the sundews that some had extolled as insectivorous vineyard watchguards, but which "only caught one insect at a time." Springing up amid vines between May and July, the foliage of *Madia sativa* would cast cooling shadows over the soil to discomfort the warmth-loving aphid underground. More to the point, the sticky globules on its stems and leaves would act as a "veritable sheet of death on which the airborne insect must find its tomb."* If the vine-growers of France wanted vengeance on their nemesis, tarweed was the stuff. Mayors should consider making its planting compulsory, said tarweed's champion. "It never releases its grip and holding its captives enchained by feet, wings or head, it lets them struggle in slow agony until they die." There was loud applause.

Louis Pasteur spoke next. France's most famous natural

---

* Maudit had planted some samples in Lyon's municipal park for the attention of delegates – but he noted sadly that the season had passed for its fatal effects to be properly observed. No insects were ensnared. It could nevertheless be fed afterwards as winter forage to flocks of sheep.

scientist admitted he was "reluctant to speak on a subject which I have not studied in the least bit . . . but what I have just heard suggests an idea, and I ask your permission to communicate it." Could not one try to find an enemy of the phylloxera?

Seven years earlier Pasteur had investigated a disease of the silkworm known as *le pébrine* that had wiped out the silk industry in southern France. He isolated and defeated a generation-crossing parasitic pathogen in eggs, silk-producing caterpillars and moths. His proposal to combat the vine-pest was a curious projection of that famous success: "Let us imagine that experiment could determine the time of year when the *puceron* comes out of the ground," he told the conference. "Then we might put a certain quantity of silkworms affected with the *pébrine* in a glass of water. By pouring this glass of water around the infected vine we might thus infect the insect. If we could find the means to infect females they would communicate the fatal disorder to succeeding generations . . ."

This fascinating proposal, with its promise of an answer from the then barely understood science of "genetics," was complete nonsense, but it was greeted deferentially nevertheless. "I thank M. Pasteur for having given us the idea. We shall try it very soon," replied Lichtenstein. "The same thought had occurred to some of us from the beginning and we recalled an American proverb – 'To decrease the number of rats, it is necessary to increase the number of cats.'"

Lichtenstein told the conference he had been in touch with Mr. Riley of Missouri, who had been experimenting with a "cannibal phylloxera" which seemed to like nothing better than to gobble up the aphids – at least those found on the leaves of American vines. The entomologist had sent samples across the Atlantic, but sadly they had all arrived dead. Meanwhile Lichtenstein would remind delegates that whatever might be found in nature to prey on the insect in France, it would have to do so in twisting crevices deep underground. The hunt for the "cannibal phylloxera" would prove just as tortuous.

18

The Ministry of Agriculture's prize remained unclaimed. There was, however, a faster way to make money. From very soon after the first formal announcement of the insect's discovery there was an outpouring of pamphlets, articles and advertisements selling snake-oil cures to despairing farmers.

What for example was the principle behind the *pestivore au moyen physico-tactique pour arrêter le phylloxéra*, or the *vinipare, destructeur du phylloxéra*? A mysterious *pâté anti-phylloxérique* was touted. The truly desperate might look to *phyllonugrane, remède radical*, or something called *la mousse céleste* for salvation.

Louis Charmet, an opportunist pharmacist of Lyon, advanced a mysterious powder with both fatal and fertilizing powers – it was called simply *Insecticide-engrais*. Having in a sixteen-page sales brochure poured scorn on other remedies (including the "cannibal phylloxera"), on the last page its inventor revealed the wonder-remedy's main ingredient. It was "dried human urine" – available for collection by enlightened vine-growers at the nearest railway station at thirty-six francs per hundred kilograms.

All sorts of curious substances were beginning to trundle in the *fourgons* of the Compagnie des Chemins de Fer de Paris à Lyon et à la Méditerranée (P-L-M) towards the blighted Midi and Gironde. There was a kind of ordered science at work, even if it could only come up with palliatives rather than cures. Knowledge of insecticides was primitive. It was a question of pour something on and see what died first, the insect or the vine, both or neither.

M. le Baron Paul Thénard, chemist and vineyard proprietor, son of the famous chemist Louis Jacques Thénard, thought the solution could lie in carbon bisulphide, a volatile chemical used as an industrial solvent and to vulcanize rubber. Vaporizing at room temperature, its heavier-than-air fumes had already been found useful in killing weevils in grain stores. At the end of July 1869 the baron arrived with a wagon loaded with barrels of the stuff at Dr. Chaigneau's vineyard at Floirac, the plight of which had so animated the vine malady committee a few days before. Baron Thénard poured the heavy liquid into two trenches dug around infected vine-footings. The aphids died, but so did half the vines. The experiment was judged to be a "disaster."

In 1873 a vine-grower called Monnestier repeated the experiment near Montpellier, using much smaller doses, with some success. There was considerable press excitement, which soon turned to gloom after the aphid reappeared the following spring. When he was blown up and half poisoned in a field experiment in 1873, Louis Vialla discovered for himself that while the chemical could be an effective insect-killer, it was inflammable, explosive when mixed with air, and its fumes were toxic.*

The French physician August-Louis Delpech had recognized in the 1850s that rubber processing was the cause of bizarre psychoses occurring among French workers who manufactured condoms and balloons in small cottage industries. Carbon bisulphide was later isolated as the neurotoxic culprit. It was also

---

* According to the U.S. Environmental Protection Agency: "At very high levels, carbon bisulphide may be life-threatening because of its effects on the nervous system or heart. Exposure can be through inhalation, absorption through the skin, ingestion, or skin or eye contact. In acute poisoning, early excitation of the central nervous system resembling alcoholic intoxication occurs, followed by depression, stupor, restlessness, unconsciousness, and possible death."

expensive – 450 francs a hectare on the first application, 300 francs a time thereafter, with subsequent applications theoretically necessary once or sometimes twice a year forever. This was perhaps affordable by the grand proprietor with his prestige wines. But for a peasant farmer with a few hectares making lowlier stuff, it was an invitation to penury.

Much depended on the soil – the insecticide worked well on light, sandy soils, far less well on heavy clay. Soil moisture level was also important – according to Planchon "not too dry when the vapor would quickly escape and not too wet when it would not penetrate to all parts and kill the insect."

Manufacturing large quantities of chemicals and dispersing them across France was as much about industry as agriculture, something well understood by the philanthropic Paulin Talabot, the chairman of the Paris-Lyon-Méditerranée railway. Hauling cheap Midi wine was the core of the railway's freight business, and something must be found to replace it. Although almost eighty years old and quite blind, Talabot had lost none of the zeal that had earlier inspired his railway-building, promotion of a canal across the Suez isthmus and the commencement of a tunnel at Sangatte near Calais to link, so its investors hoped, the railtracks of France and England. That project might have stalled, but the patriotic businessman saw his duty in putting technology at the service of the nation in the fight against the aphid, after French weapons and generalship had so disastrously failed to hold off the Prussians.

The company sponsored the work of Professor A.-F. Marion of the Faculty of Sciences at Marseille. An eight-hectare site was cleared at Cap Pinède, just north of the port, where, screened by a pine forest, hundreds of vines were planted in every possible vinifera variety to act as experimental subjects. Professor Marion and his fast-growing band of assistants were afforded free first-class travel and provided with fancy engraved calling cards – *service spécial pour combattre le phylloxéra* – as they traversed the blighted districts like exterminating angels.

Insecticide factories were quickly established at Marseille, Lyon, Libourne, Mérignac and Narbonne. A special *baril* was devised to transport it – holding one hundred kilograms with a little tap to draw off the required dose of carbon bisulphide. A device was invented called a *pal*, a tank with a reinforced metal prong that could be injected into the ground like an outsize hypodermic syringe to release measured quantities of the evanescent liquid underground.

It was a curious operation, turning a railway company into an agency of chemical warfare – with its general M. de Lamolère, chief administrator of the *service phylloxérique*, its officer class of uniformed *moniteurs* and its injector-wielding infantry known as *piqueurs*. But in the circumstances, Talabot's philanthropic advocacy of carbon bisulphide was to be welcomed. At least that is what the politicians and civil servants on the Central Commission in Paris thought. The life-scientists, though, had their doubts. For now they chose to keep them to themselves. Until the first fruits of those strange new vines growing in the Midi proved if they were right or wrong, they had nothing practical to offer.

In the west of France meanwhile a new phylloxera outbreak was nibbling at a unique and hugely valuable culture – the distillation of cognac from the wines of the Charente. In 1874, Édouard Martell, producer of the eponymous *eaux-de-vie* and a member of the Chamber of Deputies, petitioned the Academy of Sciences for help. Two delegates were sent, Maxime Cornu of the School of Natural History, and Pierre Mouillefert, professor at the National School of Agriculture at Grignon.

They arrived in Cognac that autumn to test a group of potential insecticides suggested by the head of the Academy himself, the esteemed chemist Professor Dumas. The most promising among them were sulphocarbonates, which when exposed to carbon dioxide in air and soil break down to release carbon bisulphide and hydrogen bisulphide gas, both of which are injurious to the aphids in appropriate quantities. In theory the slow release of carbon bisulphide into the soil should solve the major problem presented by the evanescent compound itself – it was washed out or evaporated once its initial killing effect was exhausted.

Sulphocarbonates of potassium, sodium and barium were applied in the winter of 1874–75 to test-plots of vines in the chalky soil around Cognac. It was established during the following spring and summer that potassium sulphocarbonate worked best. The work continued the following winter. In laboratory tests, high concentrations were observed to kill the aphids in fifteen minutes. How should a *vigneron* know his foe had really been extinguished? Pierre Mouillefert offered this simple guide: "All movement ceases, its sucker is no longer stuck into the plant tissues and it can be easily picked off. The back bends and the posterior part curls back on the abdomen, its color rapidly tarnishes into a cadaverous black."

Field experiments were successful, but the process was expensive (around 350 francs per hectare) and cumbersome – without immediate, heavy rain, large quantities of water were required to get the chemical deep enough into the soil. A system of steam pumps, air compressors, dismountable pipes and rubber hoses was later devised by Professor Mouillefert and the engineer Félix Hembert, which grander growers could summon to their domain with an ensemble of skilled eradicators like a travelling circus. In 1876 Mouillefert founded a company, La Société Nationale Contre le Phylloxéra, and would find eager clients, especially in the south-west, when it came time to defend the *grands crus* of the Bordelais.

Professor Mouillefert meanwhile was admirably dispassionate in his advocacy of one chemical or another. What mattered was cost-effectiveness. In 1877 he gave readers of *L'Agriculture pratique* two fascinating lessons in viticultural home economics. The first was from M. Damaniou's vineyard at Sainte-Foy on the left bank of the river Dordogne. It had first become infected in 1874, and its production collapsed within two years. "This man of progress had tried to fight the plague with all the principal remedies that had been recommended," according to Mouillefert, who himself had suggested potassium sulphocarbonate. It was effective, but far too expensive: saving the vineyard would cost more than any wine would be worth.

Damaniou had turned to cheaper carbon bisulphide mixed with coal tar, as recommended by the Association Viticole of Libourne, using a home-made injector of his cousin's devising to get the chemical into the ground. 'With this very simple device, a man aided by a woman can make two to three thousand holes in the ground in a ten-hour day," wrote the professor.

As it requires three holes per vine, a hectare can be treated in eight to nine days. After putting the sulphur of carbon in the hole, the woman energetically bashes the ground with an iron bar. The price of this treatment is:
600 kg of a mixture of sulphur of carbon and coal tar at 24 francs per 100 kilograms 144 francs
9 days of a man at 2.25 francs per day 20.25 francs
9 days of a woman at 1.25 francs per day 11.25 francs
Total per hectare 175.50 francs

The professor's fact-finding trip took him next a few kilometers up the river to a very curious vineyard indeed. At Nougarède was a "penitentiary colony." Its inmates were offenders against the law spared the hardships of adult prisons by their youth. The *vignerons* were all aged under sixteen, "Protestants by religion – whose perverse and irreligious natures it was hoped might be bettered by moral instruction and manual work, especially in the fields," according to Mouillefert.

The colony's director, M. Rey, a Protestant pastor, cultivated twenty hectares of vines with the labor of a hundred youthful *détenus*. When there was not enough work at Nougarède, the more trusted inmates were sent out to work for the day in surrounding vineyards. In the good times the teenage labor had earned the colony eight thousand francs a year.

Now, three years since the arrival of the phylloxera, the figure had fallen dramatically. The colony's own production from two vineyards on the banks of the Dordogne had slumped from two hundred *barriques* of wine a year to twenty. Rey had tried pyrites of copper as an insecticide, and a curious chemical *préparation antiphylloxérique* made by Messrs. Hall & Co. of London. Neither had worked. The labor used to apply it, however, was presumably free.

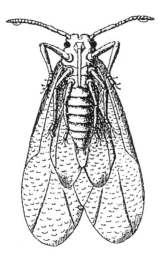

19

The richer proprietors in the south-east of France had put their faith in America. But there were many who still hoped and prayed their vines might be saved by a cure found closer to home. Prize-hopefuls bombarded the Ministry of Agriculture with surefire remedies. In the spring of 1872 the Hérault Vine Malady Committee received a government subvention to formally test some of them. An experimental plot at a vineyard called Mas Las Sorrès near Montpellier was established under the watchful eyes of Messrs. Durand and Jeannenot, professors at the School of Agriculture. It would be a long and fruitless undertaking.

To begin with it all looked quite simple. Three plots were established, each planted with twenty-five vines: Vigne Sud with old and by now thoroughly phylloxerated vines; Vigne Nord with new-planted ones; and the third, Plantier du Pins, physically separated from the others by a wall and farm buildings. The vines themselves were arranged in five-by-five-meter squares, 1.5 meters apart. Each square contained the subject vines and the control, with a post-borne placard in the center announcing the supposed remedy under test. A scale was established, zero to twelve, from complete mortality to "good vegetation." Length of shoot was a factor, as was weight of grapes harvested. The rest depended on visual observation. Other factors such as grape sugar content were built in later. It was a matter of dosing the test subject, leaving the control alone, judging the vegetative state at waypoints as the trial progressed, giving each an overall mark, subtracting one from the other – and there it was, the coefficient of efficiency of the procedure.

Between July 1872 and August 1873, 136 methods were tested. Several were proposed by commission members themselves, others came from vine-growers, amateurs, cranks and outright lunatics. There were liquids, powders and chemical smokes, animal, vegetable and mineral. Some were old favorites by now – cow's urine, copper sulphate, powdered tobacco, walnut leaves. Others were most novel. *Procède Pollier* for example proposed pouring a cocktail of whale oil and petrol onto the footings of suffering vines (score: minus one). The sovereign remedy *Procède Louis-Philippe* solemnly advised the application of hot sealing wax to pruning lesions. A mysterious *poudre particulière* was advanced in the *Procède Madame Thouret* – which unsurprisingly scored nil. There were several of these anonymous substances, as well as equally anonymous hopefuls who proposed not-so-secret remedies. "Monsieur X-X" for example advocated irrigating vines with crushed bone dissolved in suplhuric acid (score: zero).

The worst score of all was given, oddly perhaps, to carbon bisulphide – minus five. Two procedures got plus-five scores: potassium sulphide dissolved in human urine and "sulphatized colza cake," which acted by releasing sulphur dioxide, an insecticide, into the soil. The accompanying *pissoir*-gleanings presumably released enough nitrogen for a last flourish by a dying vine. Most procedures scored zero.

Planchon was characteristically disparaging. "Even with a relative success, the treated trial-squares never regained the vigor of the healthy vineyard," he wrote. "With such marginal results at such price, it would have been madness to repeat these experiments on a large scale." Charles Riley got straight to the point: "All insecticides are useless."

In July 1874 the Ministry of Agriculture upped the prize for a practical remedy to 300,000 francs. It was approved by a vote in the National Assembly. The flow of proposed cures turned into a torrent from all round the world. That September the *Scientific American* reported excitedly: "The reward we believe is open to citizens of all nations. The successful inventor will not only earn worldwide fame but a large fortune, for the sum will probably be greatly augmented by the private awards offered by the wine manufacturers of southern France." When a month later a reader suggested the copious use of cow dung – offering to split the reward with the journal's staff should his proposal be printed, the editor wrote: "Should the remedy be adopted in France as effective, we request the French authorities to remit the amount of the reward to this office, without any formalities." The Central Commission did not find things so amusing, reporting wearily a year later:

> *More than six hundred procedures have been communicated to the Ministry . . . most of them have clearly been sent by people who have no actual knowledge of the phylloxera . . . others have drawn on caprices of their imagination and have proposed incoherent mixtures of poisons, insecticides and diverse substances. Generally these communications have been sent without documentary evidence that these procedures have been subject to any actual practical test . . .*
>
> *When we see for the twentieth time the same proposal to use tobacco, petrol, seawater, coal tar etc., it does nothing to inspire confidence in these methods.*
>
> *It is not unnatural that an inventor who thinks he has found an appropriate method should wish to publicize it as soon as possible – but the prize cannot be awarded until after a convincing and sufficiently prolonged demonstration of the efficacy of the discovery and only after the most authentic experimental test will it be judged.*

The experiments would continue for many more years. In the

spring of 1876 thirty-five varieties of American vines were planted alongside French *cépages* in what was called *l'école comparé*. For some strange reason twenty more American varieties were grafted onto still-vigorous vinifera roots. They died.

By the following year, when a comprehensive book was published on the Las Sorrès experiments, almost seven hundred "procedures" had been submitted to the Ministry. They came not just from France but from all over the world – from Brazil, Boston and Birmingham, whence a certain Mr. Cope advanced the merits of douches of elder leaf tea. "The perusal of this file of nonsense casts a sad light on the state of mind of the general public," said Planchon. His Montpellier professor colleagues rejected 379 proposals outright, either because they were duplicates of already failed experiments, or because they were plain mad.

The wilder shores of the imaginings of prize-hungry hopefuls are not recorded in the Las Sorrès report of 1877. But enough comes through in the despairing scorn of professional scientists to reveal which "illusions and deceptions" were just too embarrassing to be associated with. The injurious effects (on phylloxera) of moles, ants, starlings, chickens, snails, crayfish, jellyfish, salamanders, hops, lard, mustard, nicotine, turpentine, petroleum, steam, asparagus, thyme, parsley, runner beans, Venus flytraps, sundews, tarweed, snow, exorcism, mesmerism, magnetism, galvanism and "electrical commotions" are mentioned in a voluminous literature. Marching bands, it was proposed, would drum the aphids out of their underground fastness. Volcanic ash from the excavations of Pompeii confounded the insect (two tons were sent to Marseille by an altruistic Neapolitan vine-grower to be collected free of charge by whosoever wanted it). Dynamite blasts might churn the soil for insecticide to penetrate. A mysterious "beating wheelbarrow" was suggested that would pound the soil with mechanical mallets and thus drive the invaders either into the sea or across France's truncated eastern frontier to plague the Germans instead. The prize stayed unclaimed.

There were, however, other methods of defense, devised not by far-flung eccentrics but by practical farmers caught in the path of the aphid's march. Louis Faucon, for example, owner of the vineyard at Gravéson whose nephews had playfully observed phylloxera "strolling along like good bourgeois going into a restaurant with walking sticks in their hands," had fought the infestation from the beginning with an idea of his own. In autumn 1869 he arranged for a three-kilometer channel to be cut to access a ready supply of water from the Canal des Alpines.

He constructed channels, sluices and bankings not just to irrigate his plantations, but to keep them flooded in a kind of giant paddy field for months on end. The subterranean pests would simply drown. The experiment was repeated in the winter of 1870–71. Faucon published a manual on the technique in 1874: flooding must be to a level of at least twenty-five centimeters, it stated, applied for forty to sixty days in autumn–winter. No marooned plant should be left on an unsubmerged islet lest it be a center for re-infection.*

Faucon's experiment caused great excitement. Planchon came to see for himself in August 1870, noting: "In all the parts of this vast vineyard where the level of the soil permits total submersion . . . the vegetation has clearly regained all its vigor, even on vines that seemed to be on the brink of death." The Hérault commissioners on an inspection tour in late 1873 observed that crop yield overall was 60 percent of that before the blight struck –

---

* A Dr. Siegle of Nîmes stated in the *Messager du Midi* in February 1876 that he had been the first to try the flooding technique, applying it at Forbarot in the Vaucluse from 28 July 1868 onwards. Whatever claims might be made of the government prize, being first in the phylloxera fight was a matter of fierce pride. The competing claims of ageing *viticulteurs* would rumble on into the twentieth century.

certainly better than anywhere else where an attempt had been made to fight the phylloxera by chemicals. The technique would not work on permeable soils, nor, obviously, on terraced hillsides, and there had to be practical access to huge quantities of fresh water. There was also the cost to consider (Faucon claimed he could do it for sixty francs a hectare, including the hire of a boy employed at two francs a day to work the sluices), and the need for extra fertilizer because flooding washed away soluble soil nutrients. But the results, so the commissioners concluded, were "very encouraging."

The *sulfuristes* had their powerful patrons on the Central Commission and in industry; the "submersionists" had found their champion in Faucon. He would spread the waterlogged word in *Le Messager agricole*, and later in a curious newspaper called *Le Viticulteur submersionniste*. Where the conditions were right, syndicates were formed to acquire the steam-powered pumps, siphons and Archimedes screws needed to flood the fields in late autumn and drain them in early spring.

In 1874 an enterprising engineer named Aristide Dumont advanced a plan to build a canal tapping the river Rhône at Condrieu, thirty kilometers south of Lyon, to bring its waters through a huge man-made delta of tributaries to the vine-growing areas of south-east France. The cost was an estimated eighty million francs. A deputy from the Ardèche proposed a funding bill in 1873 with the support of seventy fellow members of the National Assembly. Destruction of the phylloxera was its avowed primary purpose. Ferdinand de Lesseps, the entrepreneur-engineer who had triumphantly spanned the Suez isthmus, backed the project and a start was made on canalizing chunks of the Rhône and Garonne rivers.

But the government had other priorities. Dumont's grand project lingered in political limbo, ritually alluded to in the Phylloxera Commission's annual reports, smiled upon by a succession of agriculture ministers, but never delivered. By the time the insecticidal waterway was seriously raised again in the

French parliament in the late 1880s, other means of defending the Rhône valley's vines had been found.

There was another "natural" defense. Faucon had himself observed early in the fight that Grenache vines growing on a little bank of sandy soil at Graveson stayed healthy, while those on the clay around them were dying. At Aigues-Mortes close to the sea in the Gard there was a huge bank of alluvial sand planted with vines that became a five-hundred-hectare beachhead of green against the phylloxerated desert of clay soil behind it. A fortunate local *viticulteur*, Charles Bayle, investigated the phenomenon and became a kind of "Faucon of the sands," widely propagating the idea. It seemed more straightforward than repeated flooding or chemical dosing, and Bayle was widely listened to in the blighted Midi.

Others came forward with further suggestions. Sylvain Espitalier, of Mas du Roy in the Camargue, dug out the infected earth and replaced it with a protective saucer of sand laced with large quantities of guano. He was lucky: his vines were planted on a dried-out mud-bowl, and nearby dunes denuded of anchoring vegetation had obligingly blown mountains of sand into his vineyard. The sheer weight of such an operation seemed to rule it out elsewhere, although it was tried.

How much simpler to move cultivation to the coast and start again – even if there was a three-year wait before a single grape might be harvested. From the mid-1870s there was a dune-rush. Those proprietors who could afford it simply moved south, taking their workers with them. In some places along the Mediterranean littoral the price of previously worthless dunes multiplied a hundredfold. The sale price of phylloxerated vineyards inland collapsed. The big joint-stock Compagnie des Salins du Midi was set up in 1876 to exploit the wonder-soil of Aigues-Mortes and of Villeroy in the Hérault. Chemical fertilizers were employed, and a special technique called *enjoncage*

– reed mats planted as windbreaks between the rows of vines – devised to stop the sand from being "stolen" by the wind. It worked then, and it works today – although it is now perhaps more profitable to harvest the revenues from camping sites than from vines amid the Mediterranean dunes.

Why did phylloxera hate sand? No one really knew, and a completely definitive scientific answer has never been found. Some argued that it was marine saltiness, but the freshwater alluvial sandbanks of the Rhône delta proved just as inhospitable to the aphid. In 1879 the chemical partisan Professor Marion of Marseille placed phylloxerated roots in a trench packed with the killer sand of Aigues-Mortes. The insects died, because of an "insecticidal action of an indefinite nature," according to the professor. Two years later an Italian scientist, Signor V. Vanuccini, seconded by his government to work at Montpellier, demonstrated a possible reason in a simple experiment. He confined phylloxera in two boxes of sand for eight days, irrigating one and leaving the other alone as a control. His results suggested that it was the simple permeability to water of the fine grains that killed the pests, drowning the aphids and the eggs. The same year M. Saint-André, *chef de laboratoire* at the Montpellier School of Agriculture, proposed a more subtle reason – that the capillary action of fine sand drawing both surface and groundwater into the root-ball was what killed the underground suckers.

He seems to have been right. The fineness of the sand was critical. Silica was best of all, calcium (powdered seashell) sand not so effective. And its purity was all-important – above just 3 percent of clay in the mix, and the phylloxera would survive. Not only did silica sand drown them when there was enough rain and ground water; the fine grains effectively imprisoned the root-suckers underground after the initial infection. The clayey soils in which the aphids prospered cracked open in summer, allowing wanderers to search for new roots.

* * *

But sand and flood worked only at the margins, while chemicals offered a palliative, not final extinction. The full life-history of the aphid was still a presiding mystery. American vines now growing to maturity in the soil of the Midi would soon begin to bear fruit. Would they really make wine that was drinkable? Some of them were looking as if they would bear nothing at all.

In the summer of 1873 Professor Planchon was commissioned anew by the Hérault committee, backed by a Ministry of Agriculture subvention, to try to roll back the continuing catastrophe. The botanist packed for a long foreign trip. Letters of introduction were obtained. Specimen jars, field microscopes and notebooks went into his steamer trunk. On 12 August he left Montpellier to take passage from Le Havre. If the answer was anywhere, it was in America.

20

The botanist arrived in New York harbor aboard the Messageries Maritimes steamer *Saint-Laurent* on 29 August 1873. Charles Riley interrupted an insect-hunting field trip to the Rocky Mountains to meet him at the quayside. The late summer days were hot, the task pressing, and Riley had organized an energetic itinerary.

First stop was Ridgewood, New Jersey, for consultations with the noted nurseryman Andrew Fuller. On their way to inspect his collection of vines, Planchon observed a rolling wooded landscape that "could have been Normandy." The wild grape Vitis aestivalis grew abundantly, some just coming to full ripeness. Pausing to consume these wayside delicacies, he was pleased to note "the absence of the taste of cassis."

Next stop was Philadelphia, where the uniform rows of red-brick houses displeased the Frenchman. Although their interior cleanliness was "extraordinary," he found the overall effect "puritanical." There was a happier encounter at nearby Germantown, where the local nurseryman Thomas Meehan turned out to have been "a simple boy gardener at Kew" whom Planchon had known thirty years earlier. There he was shown a "black sweet water grape" (a vinifera variety, evidently) grafted onto the roots of a wild species from Texas known as "the Mustang." This was the only way to get a European vine to grow in the soil of Pennsylvania, Planchon was informed. When Riley returned to his entomological duties in St. Louis, Planchon ventured on alone, his Kew-learned English serving him well.

On to Washington, D.C., the capital, to inspect the botanic gardens maintained by the U.S. Department of Agriculture in a splendid greenhouse and the vines of the Experimental Garden of the Smithsonian Institution. There, knowing well what to look for, Planchon found a small quantity of phylloxera leaf-galls. With the completeness and order of the D.C. collections he was able, by comparing plant health and degree of root infestation, to discern, in his words, the "principal question of the program – what varieties had the most resistance to phylloxera."

The visitor was inspired by views of the Capitol, but observed that the principal street running to it was lined with grocers' shops which had an "unpleasant lingering odor." The city was broiling hot. It was time, Planchon felt, "to find the reviving atmosphere of woods and fields."

He found just that at the next stop, at Ridgway, North Carolina, in the "rustic home" of J.-L. Labiaux, a Belgian ("but French in his sympathies") who in spite of dire warnings had two years earlier planted a huge quantity of Aramon and other European varieties supplied by Gaston Bazille of Montpellier. They seemed to be healthy enough, in spite of the wild labrusca and riparia the botanist noticed lurking in nearby woods. Labiaux's hopeful plantation was doomed, Planchon thought, ascribing the survival of his vinifera so far to the sandiness of the soil.

Westwards to Cincinnati, capital of American wine-making, riding the Baltimore–Ohio railroad. The provincial botanist took to American trains with schoolboy enthusiasm. He marvelled at the comfort of the saloon car with its ornate gothic-style windows, luxurious plush seats and iced-water dispensers. Newsboys gave out journals at each stop to the prosperous passengers. After Harper's Ferry the locomotive began its long climb through the Blue Ridge Mountains, the hillsides covered with "an ocean of verdure," wreathed at their passing in smoke and steam.

Planchon became quite philosophical as the chuffing wood-burner wound higher into the Alleghenies, "crossing all obstacles with American boldness – demonstrating to the amazed traveller the contrast between the power of modern man with all his mechanical devices, and the quiet force of a still half-virgin nature." Wild vines festooned the woods, he noted, with such abundance one could "almost reach out and touch them."

Cincinnati was smoky, industrial. The trees lining its "sad-looking" avenues were "shabby." He observed: "Many Germans. More vivacity on their faces than among the Yankees of the north-east." Perhaps they were cheered by the still recent victory of their fatherland over France; or by the huge bunches of grapes – Concords and Delawares – that Planchon saw being offered for sale by Italians at street corners for just fifteen cents. The locals seemed to enjoy them, although the visitor thought their taste "foxy."

Planchon sought out the Cincinnati sage of American viti-culture, Robert S. Buchanan, author of *Culture of the Grape and Wine-Making*, the industry's standard reference since it was first published in 1856. They discussed such matters as summer grape rot, a local phenomenon, and the reason for the slow decline in the productivity of Catawba plantations over the past twenty years. Planchon suspected phylloxera. His host had never heard of it. They visited Emile Werk's vineyard at Westwood, not far from the city, where the botanist found tell-tale root deformations on Catawba vines. At the cellars of Werk & Son at Middle Bass Island out on Lake Erie, Planchon inspected a novel horse-powered bottle-cleaning machine and tasted the vineyard's product. He thought it "very agreeable."

Having sailed in the "elegant little steamer *Gazelle*" to Kelleys Island, the botanist encountered the Bavarian-born Thomas Rush (his name presumably Americanized), director of the Kelleys Island Wine Company. Rush had taken over an established business in 1860 and restocked it with imported vinifera from his native southern Germany. Within three years they had perished. "Only two or three miserable traminers survived," Planchon noted.

The Kelleys Islanders had started again with local varieties, and raised a miniature Ludwig II–style turreted castle of cut stone. Smoke came out of the top. It housed a fifteen-horsepower steam engine powering clever belt-driven elevators, a destemming machine and grape-skin presses. By separating juice from skins and seeds in the first operation "foxiness" was apparently reduced to a minimum, Planchon observed. In a single day this modern miracle could process seventy-two tons of grapes, "brought from all the country around." The botanist was astonished. But was the wine any good? "To judge in detail these wines of America would be somewhat beyond my competence," he diplomatically told his hosts.* Better to organize the dispatch of bottles back to Montpellier for some proper experts to sample. They would get their chance to taste them the following year.

American wine was sold by grape variety, not vineyard or region, Planchon noted with surprise. "They are simply given the name of the vine from which they are produced – or some broad European title – 'oporto,' 'claret,' 'hock,' 'riesling.' Although the quality of the *terroir* and local conditions certainly exercise an indisputable influence on the products of this

---

* Planchon's private observations made on his American tour were less than laudatory: "Notes of tasting:
  1: Norton's Virginia. Red, complicated enough wine, the bouquet recalled Burgundy of secondary quality. Pleasant nevertheless.
  2: Sparkling Catawba, the 'American champagne' made at Lake Erie. Little or no mousse, strong amber color. Pleasant taste. The impression is more that of the blanquettes or Saint-Péray than true champagnes.
  3: Riesling made by Messrs. 'X.' White wine supposedly from California – dry, very acid; neither softness, nor bouquet, inferior.
  4: Angelica. Another wine said to be from California. Diluted syrup in which the ladies delight. No vinosity; no strength no bouquet. Very inferior.
  5: Clinton. Red, rather common wine. Little foxy taste.
  6: Taylor. Dry, pleasant white wine; delicate bouquet, hint of bitterness."

immense country ... the European notion of the name of a *côte* or particular vineyard being attached to a wine as a title of nobility has yet to penetrate."

But he could see the potential. The United States might one day, perhaps, have an industry to rival that of France: "It is not lack of space that might restrict the expansion of the vine in the United States, nor is it the number of consumers. More and more the immigrant streams from Germany increase the taste and need for wine of this diverse people. The high price of labor is the only obstacle and the only possibility of defeat lies in endemic diseases that compromise certain varieties." The figures were already impressive. Two million acres of the United States under cultivation of the vine had made fourteen million gallons of wine in 1871. The nation's output of whiskey was four times that, Isidor Bush would inform him, but Planchon was pleased to note a recent sharp fall in the distillers of Kentucky's malevolent production of inflaming liquors – "because of the greater consumption of wine."

On to St. Louis, by railroad across the "monotonous plains of Indiana and Idaho." The prototypical French nature of the city had been swamped, Planchon observed sadly, by "Germans, Irish and Yankees," whose drinking habits he found disagreeably coarse. But amid the tough-guy saloons a few old French families still "occupied the higher reaches of society," he noted cheerfully, who "would only drink imported French wine." So, reportedly, did the francophone communities of Louisiana and Canada stay faithful to the wines of their long-abandoned homeland. Enough *Mitteleuropa* influence meanwhile obtained in the state of Missouri to make domestic wine-making a booming activity for consumption by an increasingly Germanic clientele, even if its makers were of more diverse origin.

* * *

At the little town of Webster near St. Louis, for example, the botanist met Mr. J.J. Kelly – "a simple honest Irishman – proud to call himself a '*vigneron*'" – who had won several prizes for the quality of his wine. His plantations of aphid-predated Catawba looked miserable, "as bad as anything in the south of France." Next door, growing in "identical soil, cultivated with the same care," Kelly's Norton's Virginia were "magnificent," testimony to the "diversity of American vines' resistance to phylloxera." Having tasted his wine, Planchon thought Kelly's products as good as, if not better than, anything that might be made in the Midi. "The excellent quality . . . and the tasteful absence of blackcurrant recommend these vines for direct cultivation in Europe without need to graft ours onto them," he wrote. That was a bold statement.

In the hot city of brick and stone Planchon was reunited with Charles Riley. The state entomologist's office, housed in an "immense building" in St. Louis, was equipped with all the latest apparatus, Planchon remarked with envy. The laboratory was lined with glass cages confining a "miniature menagerie" which allowed his friend "to observe hour by hour the phases of evolution of the entire insect world."

With such godlike powers, Riley's quest to crack the aphid's life-cycle had not surprisingly progressed. His official report on the insect for 1872 had particularly examined the "means of contagion from one vine to another." In the U.S. phylloxera was endemic – nothing to learn there – but the pattern of its progress in France, erupting outwards from "nucleii in vineyards that had never showed signs of the disease before," surely pointed to the winged form as the culprit, and the intervention of an as-yet-to-be-discovered male.

In summer 1869 Planchon had hatched his own "elegant little *moucheron*" in a glass tank in Montpellier. Three years later Riley had managed to do the same. It was clearly an exciting moment:

*Our winged female is a reality! What then are her func-
tions? In the breeding jars she invariably flies towards the
greatest light, and her large compound eyes and ample
wings indicate that she was made for the light and air. We
have also seen that she is burdened with two or three eggs
only and it is my opinion that after meeting her mate, her
sole life duty is to fly off and consign her few eggs to
some grape-vine, and that the lice hatching from these eggs
constitute the first gall-producing mothers.*

*Imagine such swarms of egg-bearing females settling
upon the vines and depositing their eggs which give birth
to fecund females – whose progeny in a single season may
be numbered by billions – and you have a plague which
may become as blasting as the plagues of Egypt!*

Riley was almost right. But those "few eggs" would not hatch
out into the hugely prolific fundatrix mothers. In fact they
would produce something much stranger: the true sexual forms
of the phylloxera. No scientist had yet seen them.

In Riley's St. Louis laboratory in September 1873 there were
plenty of winged phylloxera in their two curiously differing
forms. He and his visitor studied them intensely. According to
Planchon:

*We followed attentively the attractions of the two forms
of the winged insect which we had brought together in a
glass box, waiting for some form of copulation – but union
was there none. And for a good reason. When studying
one of the pretended males under the microscope I observed a ripe egg in place of the central vesicle.*

Both winged forms were females.

* * *

There was another important discovery to be made in St. Louis. The city was home to the Frankfurt-born botanist Dr. George Engelmann, enthusiastic researcher for many years past into the wild vines of America. His "herbarium" of dried plant samples was exemplary. It contained something fascinating, a vine leaf gathered in Texas by the French botanist Jean-Louis Berlandier in 1834. There were galls on its underside. Engelmann conducted a miniature autopsy. Delicately he cut the galls open to reveal a collection of tiny mummified insects – they were phylloxera. Planchon was especially excited. Not only did this reveal that the aphid had been around in America at least thirty years before the first European infestation; it proved that the combative Laliman's latest hypothesis, that the phylloxera had somehow recently been imported from Europe to the United States, was utter nonsense.

There is a hint of obligatory *politesse* towards his hosts in the several accounts Planchon published of his American tour. In his unpublished notes of his travels he made some wryer observations on American society, and the place (or not) of wine within it. There were very few cafés, he lamented, a boon for public morality but not for sociability. "For the ladies there are ice-cream parlors, gentlemen go there too but these are not meeting places for conversation." When the botanist did manage to find a bar, anonymous wines were placed on the counter in "small glasses." Otherwise it was beer, hard liquor and the ubiquitous iced water, "to which the Yankees seem dedicated – but which is doubtless the cause of the universal dyspepsia which, we are told, threatens the adulthood of every American."

Perhaps hotels would serve refreshments more elegantly. These modern establishments were "clean and luxurious with marble washbasins, water closets, gas light and a telegraph bureau," Planchon recorded with astonishment. In the capacious dining rooms, waiting on tables "was often done by Negroes (especially in the south); often by boys of white race, sometimes by young girls always neatly dressed with a kind of unaffected smartness." There were disappointments, however. "One talks little at table. You don't really converse . . . and the cuisine appears poor to a Frenchman, especially the ice water which does not replace wine to advantage."

Plates were not changed between courses – something the botanic tourist found extraordinary – and hot tea was served continually in huge cups without saucers, thus "burning the fingers." Napkins were minuscule, no good for wiping anything. Dishes were unseasoned, diners being obliged to choose their own doses of salt and pepper, and there were no *ragoûts*, only steaks – which were "sometimes good." There was plentiful ice-cream for dessert, but it was "much inferior to that in France."

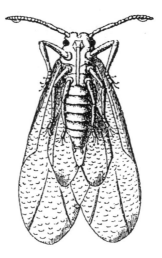

21

Professor Planchon returned to France from his transatlantic exertions in October 1873. The following year he published two long articles on the biology of phylloxera and the results of his American tour in the *Revue des deux mondes*. His extensive report *Les Vignes américaines, leur culture, leur résistance au phylloxéra et leur avenir en europe* was published in Montpellier as a popular paperback book in 1875. A line drawing of the still-mysterious winged phylloxera graced the cover. The fact that Planchon's *mission viticole* had been undertaken in response to a government commission was pumped up in the introduction, adding gravitas to findings that remained highly controversial.

The insects on both sides of the Atlantic were the same, Planchon proclaimed. There was no doubt whatsoever. American vine species showed resistance to its predations in varying degree.* Why? The answer came from Darwin, even if Planchon seemed reluctant to spell it out directly.

Given the age-old existence of phylloxera in the United States and the rapid death of European vines, it is clear

---

* Planchon proposed classifying them into three categories: "The unhurt, which are not attacked by phylloxera: they are varieties, as Scuppernong, of the species Vitis rotundifolia. The resistant, derived from the species Vitis aestivalis (Herbemont, Jacquez, Cunningham, etc.), the species Vitis cordifolia and Vitis riparia, (Clinton) and Vitis labrusca (Concord). Non-resistant varieties (Delaware, Isabella and Catawba)."

*that American vines possess the power to resist their secu-
lar enemy. Wild vines enjoy such protection and they are
generally sturdier than their civilized descendants.*

*It is possible that a process of natural selection bit by
bit eliminated those wild vines that could not fight the
enemy insect in varying degree. These in turn have made
cultivated varieties. This hypothesis proposed distantly by
Mr. Riley would explain the resistance of certain American
species, the half-resistance of some and the relative vulnerability of others.*

In fact Charles Riley seems to have embraced an evolutionary
explanation outright. The resistance of American vines, he
wrote in his State Entomologist's Report for 1871, might
readily be understood by "those who rightly believe the Darwinian hypothesis of development, that life is slowly undergo-
ing modification just as it ever has since it had an existence
on earth."

The insect's astonishing change of life from leaves to roots
could similarly be explained. "First we might expect, and those
who believe in the Darwinian hypotheses certainly would,"
Riley wrote, "that presuming our insect to have been imported
into Europe, it would have undergone some moderation of habits, not only because of its change of climate – but having to
live on a different species – Vitis vinifera. Hence its normal
habits there of feeding on the roots have been gradually acquired."

Riley was writing for God-fearing Missouri farmers. When
it came to insects, Mr. Darwin seemed OK by them. Planchon
was writing for Paris intellectuals, among whom evolutionary
theory was still getting a bumpy ride. But at last the botanist
got to it: "It is the battle of life, where the strong resist, the
feeble succumb, or where the strong survive only after the gradual destruction of the weak," he wrote in the *Revue des deux
mondes* in 1874.

The theory was perfect. The question was what to do next. The vines of the Midi, of Bordeaux and Burgundy, indeed the whole of France, if that was what it came to, could surely not be supplanted entirely by Americans, as the enthusiastic Laliman was by now loudly proposing as a prophylactic measure in advance of the aphid's otherwise inevitable triumph. To plant in as yet unblighted vineyards would just spread the infection faster, Planchon argued. There was another and much more important question. Would American grapes make wine that any Frenchman would want to drink?

Léo Laliman had no doubt that they would. In November 1873 he sent a mixed case of various vintages made from the fruits of his American vines to Louis Pasteur, ostensibly to see whether the famous scientist's new technique of heating wines the better to conserve them would work with those made from his *cépages américains*. It was a curious gesture. It seemed to be more of a publicity stunt – if so, it had unfortunate results.

"I did not regard myself as competent to judge the quality of these wines," Pasteur told the Academy of Sciences, "so I left them abandoned in a cellar until a favorable moment." That came with the arrival of M. Bouchardat, wine expert of the Société Central d'Agriculture, who "came to my laboratory in order to appreciate the wines of M. Laliman." It was not to be a pleasant experience.

Bouchardat thought the '72 Clinton equivalent to "a good *vin ordinaire*." The '71 Herbemont was "strange." The '73 Delaware was "acid – with a strange after-taste – no good for sale in France." The '68 Isabella he judged "bitter, very strange and disagreeable."

Professor Planchon clearly had similar reservations about the wines he had sampled on his American tour. Although he did not share "the prejudice which regards any wine of the United States as undrinkable," he wrote in summer 1874, "it does not enter my head to advise the wholesale replacement of the vines that are the wealth and glory of France." But *vin ordinaire* was a different matter: "As for ordinary wines . . . which are also a source of wealth for the producer and prosperity for the consumer," he wrote, "one can wonder if the direct culture of some American vines would not be the fastest and the surest way to repopulate temporarily the vast spaces where the phylloxera has brought ruin."

A glance at Duclaux's invasion maps, published in his reports to the Academy of Sciences, showed just what Planchon meant. The red ink was spilling ever outwards. By autumn 1874 the Midi infection was a vast triangle engulfing five departments. The Gironde outbreak was spreading eastwards in the direction of the prevailing Atlantic winds. On the left bank of the great river the Médocains prayed they might yet be spared, although there was already a rogue spot near Macau. A big blotch defaced the Charente, cognac country, erupting from M. Ferrand's nursery at Crouin where American vines had been imported from Indiana seven years earlier. There were the first spots in the Loire valley.

In July 1874 the Ministry of Agriculture had circularized prefects giving them local powers to ban "the introduction of any transplant, stem or root of the vine from departments where the presence of the phylloxera had been notified." The prefects of the Rhône and Corsica went further, ordering destruction (without compensation) by burning of infected vineyards. There was uproar, and Paris quickly ordered that the edicts be rescinded. As an administrative move, however, the import ban looked quite reasonable. But without means of enforcement, it was meaningless. The aphid obeyed no laws but the direction of the wind, the navigability of the soil and the supply of fresh vine-root. If a real defense was not found soon, the wealth and glory of France were going to be obliterated anyway.

22

When Louis Pasteur had raised the question of finding a natural enemy of the phylloxera at the Lyon congress in September 1872, Charles Riley was already experimenting with such biological means of pest control. In 1871 at a site at Kirkwood, Missouri, he had observed a parasite of the Plum Curculio, a snout-beetle pest of fruit trees, doing the farmer's work of eradication for him. It was a tiny ichneumon fly called *Sigalphus curculionis*.

"However utopian my scheme may appear," Riley wrote, "I intend to breed enough and send at least a dozen to every county seat in the state and have them liberated into someone's orchard." Perhaps phylloxera had a similar natural North American enemy. The leaf-gall was a clever defense against above-ground predators, but leaf-galls in France at least were proving rarer than hens' teeth.

It was "a question of finding allies against the most redoubtable enemy, the subterranean phylloxera," Planchon wrote in 1874. A kind of insect terrier was needed to go after the aphid deep underground. Two years earlier Riley had discovered on Missouri vine-roots a minute white "acarus," unobserved previously, which seemed to be attempting to eat live phylloxera and their eggs. It was a "tyroglyph," commonplace apparently in the rinds of riper cheeses. The entomologist and the botanist had discussed the counter-attack potential of *Tyroglyphus phylloxerae* in Riley's St. Louis laboratory. Riley was by no means sure, but it was worth a try.

Planchon had borne samples of phylloxerated roots back to France from his American tour with the "good" parasite on them. Their numbers had fallen during the thirteen-day transatlantic journey, he noted, and a new life-stage form of the microscopic mite had appeared. The first cold of winter killed them, so no further experiment was possible. He would have to wait for Riley to send more. Planchon gave a discreet lecture to the Academy of Sciences. Word got out. Much to his embarrassment there was a press sensation on both sides of the English Channel. The *Times* correspondent in Boulogne reported excitedly:

> *M. Planchon, a French naturalist . . . ascertained that the native American Phylloxera is identical with the insect plague that has spread its ravages over 1,100,000 hectares of French vineyards. Finding certain American wine-growing districts in Missouri and Illinois free from that pest he investigated the cause of their immunity . . . due to the presence of an acarus, a tiny foe to the phylloxera which he preys upon . . . tracking them through their earthy passages on to the very rootlets from which they extract the life juices of the vines, destroying them under any of the Protean forms of their polymorphous life.*
>
> *Numerous specimens of the American Phylloxera cannibal have been safely imported . . . and his progress will be watched with intense anxiety by millions. How he fares will not be known until next summer . . .*

The gladiatorial contest would prove disappointing. In St. Louis, Charles Riley discovered that his laboratory tyroglyphs were not quite as voracious as he had hoped. "Since numerous newspaper articles have appeared I have had several orders from Europe for samples of the cannibal," he wrote ruefully in his State Entomologist's Report for 1873. "Professor Planchon it is true will attempt to introduce it, we may hope with success . . . but it is evident that the efforts are doomed to disappointment." When Missouri-bred tyroglyphs were eventually released in the Midi they proved completely uninterested in French phylloxera. Perhaps they preferred the local cheese. Higher animals than mites predated insects. Where phylloxera appeared above ground, perhaps birds might peck the plague to extinction. In 1874 the Bordeaux oenophile Théophile Malvezin proposed an absolute ban on hunting *petits oiseaux* "by any means other than the rifle." Proscribing nets and bird-lime would effectively stop the trade in caged songbirds and tiny avian delicacies served up by restaurateurs. Instead of themselves being eaten, the little birds of the Gironde should be allowed to gobble up harmful insects, Malvezin argued in a passionately worded pamphlet. It was a big issue in rural France, but love of *la chasse* and a passion for tiny ortolans fed on armagnac-soaked bread proved stronger than hatred of the aphid. In 1879 a bill banning hunting was moved in the National Assembly. It additionally proposed that the parents of miscreant birdnesting children should be fined one hundred francs. It was voted down.

The unseemly scramble in the Midi to get American vines into the ground had proved an expensive catastrophe. Isidor Bush's nurseries had been overwhelmed with initial orders for 400,000 cuttings to be sent across the Atlantic. Hermann Jaeger shipped three railroad carloads of Vitis rupestris from New Switzerland. With Planchon and Riley's "table of resistance" to navigate by, more than seven million cuttings of Concord had been dispatched from Missouri by 1875.

Shipping costs already made them expensive, and a secondary speculative market drove up the price. It was thus the wealthier owners who got their hands on them first. Some planted a few experimental plots, others put scores of hectares under the wonder-plants. Anything to get vines growing again.

There must be something to sell, something to drink. There was virtue it seemed in the old Provençal proverb: "For want of thrushes one must eat blackbirds."

As with pioneer motoring enthusiasts a generation later, there was an explosion of amateur excitement as the American vines began to show what they could do. Newspapers were filled with hints and tips, stories of triumph and disaster, blame and recrimination when things went wrong. Lots went wrong. What came out of the packing crates shipped from Missouri was often not what had been ordered. This was brave new territory for peasants whose language was not even French.* And when planted, while some American varieties seemed to show exemplary resistance to the aphids, others did not like the soil of the Midi. In chalky soils, it was observed, the leaves of some American vines would turn yellow due to a condition known as lime chlorosis. Death followed rapidly.

But a much more serious problem was beginning to show itself. Planchon and Riley had got their phylloxera resistance scale wrong. The much vaunted Concord (a variety of Vitis labrusca) was fatally vulnerable to phylloxera. Léo Laliman had been right about the "race Labrusca," something Professor Planchon would take years to reluctantly admit. The first rush of joy as spring foliage burgeoned in replanted vineyards gave way to despair and anger as, after two or three years, the miracle vines showed their frailty and succumbed. Félix Sahut later recalled: "Many vine-growers, having at this time only very incomplete information, chose one or two vines that they

---

* They spoke the ancient regional language of Occitan. An official 1863 inquiry reported a quarter of schoolchildren in the Hérault "speaking no French at all." Gaston Bazille recalled an encounter in June 1871 with a Midi farmer soon after an article by the Senator in praise of American vines had appeared in the local press. He was asked where he could get hold of the miraculous vines. "The English name of the variety was mangled certainly in the mouth of this *vigneron* of Villeneuve . . . but the principle was perfectly well understood. I complimented my visitor on his initiative but had to tell him that, having run out of American plants myself, I could not help him."

supposed to be the best. They planted them with high hopes of success but mostly they found only bitter disappointment."

Louis Vialla, for one, would not be downhearted. "American vines are still our best hope," he wrote in 1876, "but in spite of the many difficulties they have presented, in spite of how much time it has taken, we should not complain. Everything takes long in agriculture . . . The eyes of every vine-grower in France are on us. Let us hope that such a great undertaking will not go unrewarded."

A wine of sorts was beginning to result.

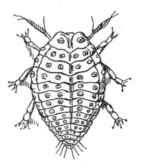

23

The Congrès Viticole held at Montpellier in October 1874 met in an atmosphere of crisis and recrimination. It got off to a bruisingly personal start. A phylloxera outbreak had been reported from Villers-Morgon, in the Beaujolais region 150 kilometers to the north. There had been a witch-hunt. The enthusiastic ampelographer Victor Pulliat (whose contribution to the phylloxera fight would later earn him a laurel-garlanded statue in the town square at Villefranche) had been experimenting with American vines for years at his vineyard at Chiroubles a few kilometers from the site of infection. He had even imported them from the same nurseryman as Léo Laliman – J.P. Berckmann of Augusta, Georgia.

Just as Laliman had proclaimed the wonder powers of American vines to the bemused grandees of Beaune three years earlier, Pulliat had extolled their virtues at the Lyon congress of 1872. Berckmann had assured him that the varieties Scuppernong and "Black-Boulace" "were the only two American races that truly resisted the phylloxera." Pulliat had then generously offered samples to his fellow *viticulteurs* amid loud applause. Now he was being treated as a pariah – the man who had brought the bug to Beaujolais.

Camille Saintpierre, director of the Montpellier School of Agriculture, came to Pulliat's defense. He himself had just returned from Chiroubles and found no trace of aphids on leaves or roots. There was no proof that Pulliat was to blame for the

150

Beaujolais outbreak. Anyway, in the circumstances, his research on American vines was entirely legitimate.
Which was the whole point of the Montpellier congress. Its
set-piece was to be a grand tasting of American wines, as open
and honest as could be. Thus far, in France at least, only scientists had enjoyed this rare experience. Planchon had sampled
them the previous summer on his transatlantic *mission viticole*,
while Bouchardat had glumly sipped Léo Laliman's wares in
the laboratory of Louis Pasteur. Now the ordinary wine-makers and consumers of the Midi could decide for themselves.
Their very future might be said to depend on the results. As
M. Leenhardt-Pomier, *rapporteur* of the Commission de Dégustation, explained: "After six years of failed efforts to find a
way to preserve or cure their vines . . . many have simply given
up trying. Rightly or wrongly they see their last hope in American vines . . . thus we have brought together as complete an
assortment of wine made from these varieties, both in France
and in America, as we can . . ."

The tasting committee had in fact done extremely well.
Ready to be sampled were French-made reds of U.S. parentage
from Léo Laliman in Bordeaux, from the vineyard of the unfortunate Borty of Pujat, from Barral at Faugio and from Camille
Saintpierre's own domain at Rochet. There were also wines
made from the experimental vines at the Montpellier School
of Agriculture. Somehow fresh bunches of grapes (if not yet
their fermented product) had been obtained for display from
a Japanese vine.

Mixed cases of reds and whites of various provenance had
been urgently dispatched across the Atlantic by Isidor Bush and
Charles Riley of St. Louis, by Messrs. Poeschel and Scherer of
Hermann, by the "Irish *vigneron*" J.J. Kelly of Webster,
Missouri, and by Professor Planchon's Belgian-born friend Mr.
Labiaux of Ridgway, North Carolina. Samples of native and
vinifera-made wines had also been sent from the Buena Vista
Wine Company of Sonoma, California, but these were thought

"of little interest as the phylloxera was not present beyond the Rocky Mountains." The jury and the public would try them nevertheless.

The sipping and swilling was over. The judges would be dispassionate. The reaction of ordinary Montpellerains, however, sampling the offerings laid out on trestle tables in the Salle Saint-Côme, had already betrayed the outcome. Much was left unconsumed.

"If the judgement of the commission, as that of the public, has been often unfavorable to the majority of these wines, it is important to observe that they have been produced in very disadvantageous conditions, especially those grown and produced in France," said Leenhardt-Pomier.

> Given a natural impatience to taste these products of American vines, and the fact there are so few of them [ready to bear grapes], one had to make do for the most part with wines made scarcely a few days ago, produced from grapes that were still green, picked from vines that were too young, harvested in such small quantities that they had to be simply fermented in their bottles . . .

The imported stuff was worse. Labiaux's North Carolina product, for example, made from white, sweet Scuppernong grapes (his vinifera had presumably died, as Planchon had predicted), was judged "hardly agreeable – the taste is almost medicinal." The rest of the offerings were evidently as grim. Leenhardt-Pomier noted diplomatically: "As for the wines produced in America, everyone knows how few vines there are as yet cultivated in that country and how wine-making remains in a primitive state." Yankee clumsiness might be excused, with all that added sugar to suit "local tastes," but the commission had asked that sugared wines be excluded in favor of "wines made naturally and simply following French procedures," and these were what they had indeed received. But no Frenchman who tried them could ignore the old problem – "foxiness."* It was to be hoped, said the judges' report, that the horrible *goût de renard* would be eliminated by "culture in our climate and our soils and by our processes of vinification." Some hope.

There was, though, a glimmer of light. "The wine which satisfied the tasting panel most is the one that Laliman exhibited, produced from his Jacquez and Herbemont grapes," said the judges. "This wine has a magnificent color and an irreproachable taste very close to that produced from the vines of Bordeaux cultivated in the same conditions." They must have been desperate.

The California wines scored reasonably enough. To complete the proceedings a viticultural curiosity was on display at the conference – a young Aramon vine grafted onto an American rootstock by Henri Bouschet of Montpellier. It appeared to be thriving. "One might count on the resistance to the phylloxera of its American roots while it would bear European fruit," Leenhardt-Pomier assured wondering delegates. On his transatlantic tour Professor Planchon had seen several such examples of vinifera grafted onto native vines. Indeed it was the only way, he had several times been informed, they might grow at all in the soil of North America. Perhaps alien roots really could take care of the dirty business underground while vinifera greeted the day.

What would their wine taste like? Just where and when the

---

* "It is impossible to define the taste of 'fox'," wrote the controversial oenologist Professor Lucien Daniel in 1910; "as with all that concerns organoleptic sensations one must try it oneself to get the idea." Why it should have earned its vulpine name is a matter for debate. In its native North America, labrusca had been known as the "fox grape" since the seventeenth century – various explanations have been offered why. Perhaps it was the taste: "The fox grape of Virginia is of a rank taste when ripe, resembling the smell of a fox," wrote Robert Beverly in *The History of the Present State of Virginia* (1705). Alternatively, "fox" meant "wild"; foxes ate them; the leaf "looked like a fox's paw." Labrusca grapes contain the ester methyl anthranilate, thought to be responsible for *le goût foxé*.

first draught of French wine from grafted vines was apprehensively sipped is impossible to determine from the literature. But the golden moment had indeed come – even if the pioneers of the Midi were as yet reluctant to subject their strange new liquors to formal judgement. The news however was spreading. The former senator Jean-Henri Fabre of Montpellier informed sympathetic politicians in Paris in November 1876: "For four years past I have grafted Aramons on American varieties and on these I have further superimposed by a second graft varieties from the Bordelais. The roots of the bordelais and the Aramons have been completely destroyed by the phylloxera while the American roots stayed perfectly healthy. In these plants so grafted the mixture of sap has been as complete as possible – yet it has produced no alteration in the quality of taste of the wine nor had any influence on the [resistant] constitution of the roots." It seemed a miracle – but it was true.

The genus Vitis, it has been noted, is biologically unusual in that its species do not fit the classic definition of a breeding group. They can be hybridized by cross-pollinating* to make an interspecific seedling. Different species furthermore can be grafted to produce not a hybrid but a compound organism. In the words of the French ampelographer Pierre Galet: "Rootstock and scion maintain their own biological rhythm . . . it is now known that rootstocks have no effect on the scion, aside from the question of vigor which can affect wine quality. For example grafting Chasselas onto Vitis labrusca or its descendants does not lead to the production of foxy grapes. Similarly rootstocks that are resistant to mildew do not unfortunately transmit their resistance to the foliage of the scion."

---

* Traditionally done by transferring male pollen to the receptive stigma of hermaphroditic "mother" varieties from which the male anthers have been removed – and then germinating seed from the resulting fruit.

The technique of grafting had been known in viticulture since ancient times. It might be used to change one variety of vinifera for another, for example, without uprooting the whole vineyard – but its practice remained a localized curiosity. Jules Guyot for example wrote in 1865: "Grafting, although admissible in the production of table grapes, should never be attempted, except in the case of absolute necessity, for wine grapes." Splicing one plant to another was not big science. Grafting was something that vine-growers could experiment with for themselves. How it worked on the microbiological level would later be an avenue of research for twentieth-century science, but the point was that it did work. To begin with it was a matter of "field grafting": the rootstock was planted, allowed to grow for a year, then in the early spring literally beheaded a few centimeters above the ground for the budded scion of a vinifera variety to be splinted into it. The compound plant was then packed round with earth to protect it from late frosts. The technique worked well in the warmer south.

Gaston Bazille explained how easy it was: "For the past three years I have grafted many thousands of young American vines," he told readers of *L'Agriculture pratique.* "For the work I simply employed laborers from a neighboring village, intelligent enough certainly, but many of whom had never used even a simple knife to do the work of precisely splitting down the middle a young stalk no bigger than the width of a finger. But they quickly get the hang of it and my grafts have succeeded well."

For a successful graft a precise yet robust join was all-important, wrapped round with "raffia" (just what that was was a huge mystery at first in peasant France) and anointed with a dab of paraffin wax. Skilled grafters could do it without the need for binding. The merits of various techniques began to be proclaimed from the mid-1870s in an outpouring of exhibitions, articles and how-to-do-it pamphlets. The lawyer-*viticulteur* Léon Gachassin-Lafite of Bordeaux, for example, in the summer of 1874 published "La Rhizoplastie – a practical method for saving vines from the phylloxera at minimum cost," which advocated simple field grafting. The same year Henri Bouschet of Montpellier proposed a curious double-rooted graft, French and American vines conjoined above the ground like Siamese twins. There was a lot of talk in the early days of "centaurs'" and "chimeras."

Bouschet showed a grafted Aramon at the Agricultural Congress held in Bordeaux in December 1875. "The Gironde will not abandon its Cabernet, its sauvignons, for American vines any more than Burgundy will abandon its Pinots and even its Gamays," he told delegates. Quite so. "It is only as root-stocks that the phylloxera-resistant American vines can perform their greatest service for us." He was playing to the right audience. A number of bold Girondins had followed the lead of Léo Laliman and planted Americans. Their fortunes, like the taste of their wines, were mixed. Grafting however would become a Girondin speciality. The formidable Mme. Ponsot of Pomerol would write a famous book about it, *De la reconstitution et du greffage des vignes*, and established a thriving nursery.

In the Gironde, M.G. Cazeaux-Cazalet devised the famous "Cadillac graft," named not for an automobile but for his distinguished wine-growing canton, a one- or two-budded wedge-shaped scion inserted diagonally into the side of the rootstock's stem. In this technique the rootstock was allowed to retain some of its foliage in the early stages of growth. If the scion failed to take, the root was useable again.

Slowly at first, then ever more rapidly, grafting skills would spread. The Montpellier School of Agriculture set up special courses. On the first day, 3 March 1879, "its lecture halls were found far too small for the crowd who wanted to attend, the register bore more than nine hundred signatures." Certificates of competence were awarded by agricultural societies, to be reverentially framed and displayed in the parlor with the same pride as an honorable discharge from military service in the defense of the nation. From 1878 the technique began to become industrialized; machines were devised to make the precise cuts for so-called "bench-grafting" capable of churning out hundreds of thousands of grafted plants in a spring season from still-green rootstocks and scions raised from cuttings in phylloxera-free soil.*

Who was the first to propose grafting as the first line of defense would later become controversial. Gaston Bazille had tried unsuccessfully to transplant vinifera onto a Virginia creeper in 1870. Léo Laliman insisted he was first to advocate conjoining Europeans with Americans:

> In the Messager du Midi of Montpellier of 7 October and 12 November 1869, I indicated resistant vines and recommended using some of them as rootstock. I find the proof of this in the newspaper Les Vignes américaines of 1877, in which the eminent botanist M. Durieu of Maisonneuve wrote that I had resolved an important problem producing Chasselas and Malbec [table grapes presumably] so true of taste that it was as if they had been picked on their own roots, whereas they had been obtained, eighteen years ago now, by transplanting them onto the roots of an American labrusca.

Foolish M. Laliman. He could not resist bursting into print or turning up at viticultural congresses to proclaim he was first to import American vines, first to produce wine from them, first to realize their resistance to phylloxera, first to graft them. So blossomed his reputation as "the Attila of France's vines."

---

* The Ministry of Agriculture would later issue strict rules: "The nursery must be established in healthy soil, never having been previously planted with vines and isolated from any other vines, preferably enclosed by a wall. Cuttings must be free of any trace of soil and rigorously washed before planting or transport in a solution of potassium sulphocarbonate."

Plenty of questions remained. Was a truly scientific scale of rootstock resistance and soil affinity attainable? Would the now-being-spoken-of "hybrid vines," that is vinifera varieties artificially crossed with Americans, survive the pest? Would grafted plants last as long productively as "own-root" plants? Would they retain their resistance to the aphid year on year? Planchon, for now, remained scientifically dispassionate. "[Grafting] is a question yet to be studied," he wrote in 1872, "on which it would be unwise to anticipate the results and offer hopes which could turn out to be illusions." Most important of all, could their wine really be as good as that which had gone before?

In the devastated Midi they had nothing to lose. To those outside the circle of destruction, grafting proud French vines onto alien roots seemed the counsel of miscegenating madmen.

24

Almost a decade after the aphid's first appearance in France the defense was still woefully ragged. The *submersionnistes* could operate only at the margins. The *sulfuristes*, for now, spoke only to the wealthy. The sand-men would make their own luck. Following the prescriptions of the *américainistes*, it seemed, resulted in either outright failure or a wine wreathed in shame.

The proprietors of the as yet unblighted regions of France (and their political and newspaper allies in Paris) watched the slow-motion disaster unfolding in the south with a sense of chilly denial. The harvests in the Médoc and the Côte d'Or were excellent. Insecticide would deliver them from evil when and if their time came.

If the pest had indeed come from America, as everyone except Léo Laliman now seemed to believe, then the import and dissemination of all foreign vines must surely be halted. The legislation of July 1874 had given prefects local powers to do just that. Newspapers were whipping up indignation against the illegal immigrants and their traffickers. It was being argued in the National Assembly and Senate that much tougher national powers were necessary. The plague carriers must be banned if the rest of France was to be saved.

Planchon despaired. In 1876 he fired his fiercest polemical broadside yet: "What's the use of looking in Europe for help that only America can deliver?" he wrote in the *Revue des deux mondes*. "Those detractors of the vines of the United States – to listen to them you might think everything is lost. If these foreign-

159

ers should invade the holy ground of the Champagne, of Burgundy and of Bordeaux – Oh desecration! These glorious products warmed by the French sun will be sacrificed to hideous
brews tasting of blackcurrant or bedbug. 'The barbarians are at
the gates of Bercy,' they cry, and the unfortunate Parisians . . .
already being poisoned slowly by fuchsine are going to be poisoned more rapidly by *gros bleus* or some vile transatlantic *piquette*."*

The professor had a problem. However brilliant his journalism (articulating the sex-life of aphids to influence the political
mood in Paris required an exceptional talent), the argument
hung on an exquisite intangible. It was in the tastebuds of
Frenchmen. American wine was dreadful, even Planchon se-
cretly admitted it. As much as he might argue that those exotic
vines now coming to maturity around Montpellier would produce wine "as good as if not better than the ordinary wines
of the Midi," the taste of fox was always going to bring the
Americanists crashing down.

In his arguments, therefore, Planchon perforce alluded to the
untried promise of grafting: "Good God! The earth belongs
to mankind. What childishness to make the choice of vines
(especially those destined to serve as rootstocks) a question of
patriotic jealousy . . . It goes without saying that while the
power of the rootstock directly influences the development of
the transplant, the rootstock does not transmit the particular
taste which it would have in its own grapes," he wrote with
newfound confidence. "The scion's natural qualities are completely unchanged. The *goût foxé* will never pass in the least
degree to the fruits of vines grafted onto these foreign '*nour-
rices*' [which translates picturesquely as "wetnurses"]." That
was his opinion. There was a lot of convincing still to do.

* * *

* Bercy: the French capital's great victualling entrepôt – wine terminal of the
P-L-M railway, and by 1870 the largest wine-trading center in Europe. *Gros
bleu*: Parisian slang for the cheapest red wine served in the lowest bar.

There was a simple-looking counterproposition. Why not re-plant with vinifera vines and keep them dosed with insecticide? Several proprietors in the Midi had tried it, and their travails were taken up by François Rohart, inventor of patent wooden blocks (cubes of spruce doused in carbon bisulphide) which when buried in vine-footings proved an efficient enough alter-native to soil injection. "From small ideas come great things," he wrote in *L'Agriculture pratique*, "a happy precedent in the face of an inextricable chaos of ideas and baroque proposi-tions." A little movement began – Le Congrès des Vignes Fran-çaises – headed by a senator, no less, M. Guyot-Lavaline, with a manifesto slamming the "innumerable theories, substances and procedures hailed as infallible in triumphing over the plague – which had only led to ruin." Convinced that "our ancient varieties and old French vines can yet be defended," the Congrès held a conference at Clermont-Ferrand in August 1879. Insecticide, submersion and the "cannibal phylloxera" were on the agenda; any mention of American vines was for-bidden.

There was still a wonder cure that science might deliver. After the first flurry of entomological discovery in 1869–71, the race to fully describe the phylloxera's life-cycle had stalled. No male of the species had yet been found. A necessary injection of "male vigor" suited prevailing phallocentric views of human society, let alone the brutish world of plants and insects. Every-thing that was known of other aphids said there had to be a vastator.

Professor Georges Balbiani of the Academy of France turned for possible answers to the study of the *Phylloxera quercus*, the aphid that infected the leaves of oak trees. While there was no root-living form, the production of leaf-galls offered promis-ing clues. After ten years of research Balbiani would eventually describe a life-cycle almost as baroque as that of the vine phyl-loxera – including parthenogenetic reproduction on the leaves, production of sexual forms, and a single egg laid in the autumn in bark crevices on the aphid's primary host, *Quercus coccifiera*, the evergreen Kermes oak. From this a female "fundatrix" arose which pricked the leaves of the same tree. Her offspring grew wings and migrated to a different species, the Sessile oak or *Quercus pubescens*. Their descendants in turn grew wings and eventually returned to the Kermes oak where they laid eggs.

In the autumn of 1873 Professor Balbiani crucially discovered that these oak-infesters hatched into both a wingless female equipped with ovaries but devoid of means of feeding, and an equally anorexic male equivalent which sported a primitive *pénis*. Charles Riley published the findings for the attention of Missouri farmers. "The sole aim of their existence is the reproduction of the species," he wrote. "They crawl actively about and gather in the crevices which are afforded them. The male, except in size, seems to differ from the female only in having a small conical turbercule which serves as a sexual organ. Coitus takes but a few minutes and the same male may serve several females." After mating the fertilized female laid a solitary egg.

It seemed plain that the vastatrix must have the same sexual forms as the oak aphid. The entomologist Maxime Cornu of the Natural History Museum in Paris found two tiny female vine-living phylloxera very soon afterwards. They were most curious: as well as having no evident means of feeding, their abdominal cavities were almost entirely engorged by a single egg. The studies continued in a Montpellier laboratory. In August 1874 a similarly wingless infant emerged from an egg laid by a captive winged female. It appeared to go through its larval moults very rapidly. It was male.

The following summer M. Boiteau, veterinarian of Villegouge in the Gironde and a keen amateur entomologist, observed just where the two slightly differing winged forms, one bearing the male, the other bearing the female, deposited their eggs on the undersides of vine leaves. They inserted their proboscis, sucked leaf-sap for twenty-four hours, laid their eggs and died. The vet gathered enough living specimens for a successful reconstruction of the eclosion of the sexual forms by winged females in his well-equipped rural laboratory. In September 1875 Professor Balbiani arrived excitedly at M. Boiteau's *cabinet de travail* at Villegouge to observe the subsequent microscopic couplings.

The romance of the sexual forms was short. They seemed to bypass larval adolescence completely. "From the very moment of hatching, the male displays numerous mature spermatic strands," wrote Balbiani. "These tiny males, so imperfectly formed for life as individuals, live only to reproduce and are burning in their ardor – being able to fertilize two females in the space of some minutes without breaking off to rest." This was to compensate for the lower number of males, he thought. The large number of females who failed to find a mate laid sterile eggs and promptly died. Successful coitus resulted in the female laying a single fertilized egg, clearly designed, thought Balbiani, to survive the winter cold and reproduce the species. He called it *l'oeuf d'hiver* – "the winter egg."

The question remained – outside the laboratory, where was it laid? Was it above ground, or directly onto the roots?

Professor Balbiani had been intently studying root-living female phylloxera, especially the pattern of their parthenogenetic reproduction. It appeared that as one generation followed another in the course of spring and summer, there was a gradual falling-off of their oviparous output. For the race to survive, he concluded, this *dégénérescence* had to be reanimated by the intervention of the newly discovered sexual forms. The male-fertilized winter egg was the newly super-fecund point of departure. Find and destroy the eggs, and the teeming phylloxera would simply die out. Balbiani published his theory in summer 1875.

It was highly controversial. The *dégénérescence* theory was nonsense, said Jules Lichtenstein. He and other investigators had been working with phylloxera for years, and had observed parthenogenetic females reproducing generation by generation through spring and summer, a tiny proportion surviving as underground hibernants to wake again in the spring. They had noticed no such decline in reproductive power.

"The myriads of sucking insects buried under the ground are the most dangerous enemy," Planchon argued, "in the sense of causing an immediate extension of the evil around the first infection. Swarms of winged insects no doubt can establish distant colonies, but once established, the descendants of these colonists will be able to multiply by themselves without needing any revitalization by so-called winter eggs." He had already concluded that root-form crawlers, once they had come to the surface, might themselves be blown long distances by the wind.

But where could the "winter egg" be found? The diligent Boiteau was on the case. Through the early spring of 1876 he stalked the vineyards around Libourne in the Gironde, magnifying lens in hand. On a brisk morning in late April he found a tiny crack on the woody stem of a two-year-old vine. It was a "minute gallery one tenth of a millimeter wide," into which an egg-engorged female had managed to squeeze before laying her precious charge and dying. On 19 April it hatched. Boiteau observed a tiny wingless insect emerge from the bark crevice, more compact and rounder than the root form. She looked like "a little tortoise." The fundatrix nymph began her six-legged upward climb to find a newly budding leaf wherein to entomb herself, moult to adulthood and lay her mountain of eggs.

There was great excitement when Boiteau published the news. If he and Professor Balbiani were right, the phylloxera had a fatal weak point, which might be attacked above the ground. The amateur entomologist continued his searches, finding several more eggs in Girondin vineyards, the provincial press greeting each discovery as if a great detective was on the trail of a serial killer. The case of the winter egg dominated the reports of the Academy of Sciences. Every vine-grower in the

Midi it seemed was looking for the fabled yellow speck. No search was rewarded. Nor could Charles Riley find it on St. Louis vines. Professor Marion, meanwhile, found one only at the P-L-M research plot at La Pinède, north of Marseille.

Although Jules Lichtenstein and the young entomologist Valéry Mayet managed to experimentally obtain several winter eggs from sexual females at the School of Agriculture at Montpellier, it was nowhere to be found in the open field. The Montpellier team went to Libourne with Boiteau as their guide. No eggs. Boiteau came to the Hérault to search, this time with the minister of agriculture himself in tow as an observer. No eggs to be found there either. The hunt went on.

Leaf-galls were proving equally elusive. Although they had been observed "frequently enough" in the Bordelais – on some vinifera as well as American varieties – hardly anyone had seen them in the Midi, at least since Planchon and Lichtenstein had found them on the mysterious "Tinto" (a vine whose origin the botanist was now describing as "indefinite") in the Vaucluse eight years earlier. Maxime Cornu had long since regarded the discovery as suspect, believing that "Tinto" was in fact an American vine. "Other than on this unique occasion no one has ever seen [galls] anywhere within the huge perimeter (more than a million hectares) invaded by the parasite in the Midi," he wrote. "Nor has it ever been seen in the south-west outside the property of M. Laliman at Bordeaux."*

A new intellectual battlefront was opening. It was becoming clear to Planchon that the aphids' adaptation to the vines of Europe was far more remarkable than had yet been realized. In the dry heat of the Midi they had dispensed with their above-

---

* In 1950 the zoologist Pierre Maillet examined eighty-year-old dried leaves from the phylloxera era preserved in the collection of the Montpellier School of Agriculture. He could identify only three fundatrix-formed primary leaf-galls on hundreds of samples of vinifera. There were more galls on European leaves, but he concluded these were "secondary infections," where leaf aphids had migrated from physically adjoining American vines on wires and trellises.

ground existence entirely. Leaf-galls occurred in the cooler, wetter Atlantic-blown south-west, but they were extremely rare and transitory.

The "winter egg" would turn out to be the snark of the vineyard. A little industry grew up around the hunt for it and the means of its destruction, including patent insecticidal distempers of heavy oil, lime and mothball to be applied to vine-stems, supposedly to kill the egg or catch the fundatrix on her short upward march. A certain M. Sabaté of Cadarsac in the Gironde devised a curious chain-mail glove with which to scrape vine-stems "clean." M. Bourbon of Perpignan invented the *pyrophore*, a miniature flame-thrower burning an aerated spray of petroleum to scorch the stems of unfortunate vines.

The *oeuf d'hiver* had many adherents in the Bordelais, where leaf-galls were more common. When it was observed that vines scraped or daubed in the winter by the prescribed methods were indeed clear of galls the following spring, there was general rejoicing – but it very quickly became clear that the root-living aphids were still feasting and propagating underground.

Professor Balbiani and his partisans never gave up. Winter eggers went on hunting for a decade and more. Planchon was as polite as he could be in quietly disparaging their efforts. As he told the Botanical Congress of Anvers in 1885: "With all the deference due to my *confrère* M. Balbiani, I can only express the deepest reservations on employing a procedure which rests on highly contentious theories – which if it were to be applied wholesale would mean distempering millions of vines on which it would have been necessary to first establish the presence of the winter egg."*

---

* So great was Balbiani's reputation as the man who had closed the life-cycle of the aphid that it took another thirty years for his winter-egg-degenerescence theory to be formally overturned. It was at last done in a 1915 paper by the Italian entomologist and hero of early-twentieth-century phylloxera research, Professor Battista Grassi.

Nor had the miraculous predator, the "cannibal phylloxera," completely gone away. In September 1877 Léo Laliman sent the Academy of Sciences a mysterious box with a covering letter. The package contained "a larva – which one might choose to consider as the cannibal of phylloxera vastatrix. It gobbles up this aphid in such quantities that I saw twenty-five of them disappear in ten minutes," he wrote. Laliman had observed the strange insect feasting on aphids in leaf-galls. He had himself set some root-living phylloxera in front of it which it had promptly devoured. "In view of its prodigious hunger, my fear is, it will arrive at the academy dead of hunger," he confided.

It was indeed quite dead. The tiny corpse was examined on behalf of the Central Commission by Georges Balbiani. He concluded it was a "Syrphus" known as a predator on certain species of aphids – but not upon the dry leaf devastator. He thought Laliman's observation was "a local accident" – and suggested he repeat the experiment in a suitable sealed jar. The Syrphus seemed to have lost its appetite.

25

France's political classes still ignored the aphid. The Paris newspaper *Le Temps* for example could still report the insect's tenth year of progress in its *La Vie à la campagne* column along with bucolic observations on prospects for *la chasse*. An interruption to the supply of cheap Midi wine was not of immediate concern to bourgeois Parisians. The industrial laboring classes were beginning to find other means of slaking their thirst. So too might rural France ignore the metropolis. "Debate over the constitution," reported the Bordeaux commander of gendarmes in 1875, might "fascinate the towns but passes almost unnoticed in the countryside – preoccupied only with the way wine is selling."

The aphid however presented certain political opportunities. Bonapartists and royalists did what they might to whip up anti-republican sentiment as the state perforce began to pry into the affairs of wine-growing Frenchmen. In 1876 Monsignor de Cabrière, the bishop of Montpellier, sanctioned devotional processions among the vineyards, not just to ward off the phylloxera but also against the "republican menace."

In fact those whom the plague had utterly dispossessed, faced with an electoral choice between "legitimists," *laissez-faire* moderate republicanism and the radical brand, tended to choose the latter, with its promise of state aid for the poor funded by taxing the rich. "The attachment to tradition was seriously undermined by the crisis," according to one twentieth-century political analysis. "A once stable population

had been uprooted . . . young men who had in the 1860s gone
to vineyards to work, now returned to their highland homes.
When replanting began they once again took the rocky trails
to the vineyards . . . still the awkward highlanders, working
side by side with local people who had lost their plots and
also with Spaniards and Italians. Under these conditions old
attachments rapidly withered. The [disease of the] wine grape
was the most important solvent of tradition."

The renewed excitements gripping the French capital, however, were man- rather than insect-made. On 16 May 1877
President MacMahon dismissed the moderate republican Prime
Minister Jules Simon. There was uproar. The administration
of his royalist successor the Duc de Broglie was short-lived –
the general election of October at last returned an overall republican majority.

Public disorder in the phylloxerated districts, however, remained minimal. The drift of France's overall rural population
from the land to the cities was turgidly slow. No one was dying
of hunger for lack of wine. A plot of land might deliver beans,
potatoes or maize to fill stomachs. According to one account
from Burgundy, "To survive a *vigneron* must become a 'culti-
vateur' [a term laden with downwardly mobile meaning], raising a pig or a cow to ensure his subsistence." But the rhythm
of life where the insect reigned was changing utterly. Slowly
the hilltops were abandoned. Peasant owners became hired
laborers.

J.-A. Barral, editor of *Le Journal de l'agriculture* and vine-
grower of Faugio in the Hérault, took a magic-lantern lecture
to Paris in an attempt to raise the capital's interest in the aphid.
There were no images of starving peasants – southern France
was not Connemara or Bengal. His audience rather was treated
to a succession of now familiar pedagogical charts of disagreeable insects and the various means devised so far to eliminate

them. When it came to describing the plight of his neighbors, however, Barral let his emotions overflow:

> *In villages where once everyone felt at ease, where the population teemed, today distress is everywhere. There is no work, so the people must emigrate to seek their fortune in the New World or in Algeria, that they might not starve in France.*
>
> *I remember some years ago, being in the district of Montpellier in September; it was the time of the* vendange, *everybody sang in the villages, a huge crowd of laborers had come down from the mountains of the Cévennes for the harvest. Everywhere you cared to look there were wagons laden down with grapes. People could work for just a few hours and earn ten or twelve francs.*
>
> *I returned to the same place two years ago at the same time of year. The houses were shuttered, there was nothing in the streets, not a single worker, no one had come down from the mountains . . . it was a scene of the most complete desolation . . .*

The fate of bourgeois *propriétaires* could be just as distressing. Barral recited the story of "a widow who had lost her husband early and who had admirably raised her daughters." She had a "big and beautiful domain" which she had wanted to enlarge and improve with new cellars and wine-presses. She had borrowed 100,000 francs to secure her daughters' future prosperity.

> *The phylloxera appeared; the vineyard declined; demands for repayment arrived, fatal! The domain which had been worth 600,000 francs, was sold for 100,000 . . . the widow and her daughters were reduced to penury.*

When the bailiff did at last arrive at the door, some had another choice. The soil of Algeria beckoned, as yet free of the hateful aphid. A trickle at first, then a flood of peasants unpicked themselves from their ruined plots, heading for the emigrant ships. Others went to Argentina and Chile, taking their wine-making skills with them.

Agriculture ministers came and went in Paris, but the argument about imposing an outright ban on foreign vines had long been smouldering. Under the legislation of July 1874 prefects in as yet uninfected wine-growing departments had imposed bans of their own, but they were virtually impossible to police. There was a rash of lawsuits as vine-growers found loopholes in local edicts against their illegal immigrants. Like Victor Pulliat in the Beaujolais, such apostles of Americanism were being treated like carriers of typhoid.

Planchon argued in the *Revue des deux mondes* in 1876 that young vine-stems arriving from across the Atlantic in winter at Le Havre, to be securely forwarded in their sealed boxes to the phylloxerated districts, would not spread the contagion in their passing. Wind and biology would inevitably do that anyway, whatever foolish humans might or might not do to hurry or retard the plague's inevitable advance.

But the government had to do something. On the recommendation of the Central Commission, on 6 March 1876 prefects in fifty-six vine-growing departments were ordered to form *comités d'étude et de vigilance*, with the aim of finding means of defense – and to "experiment with the procedures proposed to stop the progress of the malady." Representatives from unblighted regions were invited to Montpellier for a crash course in identifying the enemy. Le Vicomte C. de Meaux, minister of Agriculture and Commerce, sent out a covering circular: "The efforts of my administration have run up against inertia and generalized indifference – and in most departments have had no effective results. Today one might hope that the incessant spread of the disease will arouse spirits to the grave danger that menaces the public good." The minister would be disappointed. The response on the ground was apathetic. Where the new committees went looking for phylloxera they were met by hostility, obstruction and denial.

A reporter for the Société des Agriculteurs de France wrote in 1878: "They who know the characteristic features of the phylloxera invasion will immediately recognize the incredulity and illusions which precede it . . . There are those who simply deny it could ever happen to them, while others seem to fall asleep, holding blithely on to some magical insect – or plant – or drug – known only to them that at the last moment will miraculously deflect the plague."

Mme. Amélie de Bompart, for example, a self-confessed "strawberry-lover" of Gradignan in the Gironde, published a pamphlet in 1879 announcing a startling discovery. Wherever her favorite fruit was planted, she observed, there were no phylloxera to be found on nearby vines. This was due to a tiny spider ("un arachnide trombidion"), she insisted, that in spring lived harmlessly enough on the leaves of strawberries. In June, however, the eight-legged *dévoratrice* "disappeared underground" to chase the aphid "like a cat does a mouse."

Sterner action from the center was clearly needed. Phylloxera was now a matter of foreign affairs. Germany had imposed a border quarantine in 1875, as had Italy and the colonial authorities in Algeria. In one of the first pan-national moves in history to tackle a biological foe, in August 1877 several wine-growing countries sent delegates to the Congrès Phylloxérique Internationale held in Lausanne, Switzerland. The mood was businesslike – the plague must be tackled "by all the means that science can furnish," said Halna du Fretay, inspector general of agriculture and leader of the French delegation. There was no question, the *puceron* had come from America, he declared. It was assuredly the cause, not the effect, of the plague. But what to do about it? There were those who proposed planting "resistant American vines, but the wine they made was undrinkable" according to du Fretay's deposition. He might however admit that "grafting," although he could not recommend it, held some promise. It all remained highly experimental however – and where was the proof that wine from grafted vines could be free of the dreaded foxy taint? Nobody seemed to have tasted it yet. Experience with insecticides had been full of problems – charlatans had crowded in – but the work of the P-L-M railway showed what could be done. In chemistry lay the best hope.

Biological quarantines must be imposed, internally and across national frontiers, and governments given powers to apply "administrative treatments" where an outbreak was reported, du Fretay proposed. The greater public safety overruled the rights of the individual to pursue his own solution. The Congress agreed.

Jules Planchon was a delegate. He may have seen his *système phlloxérique* at last internationally vindicated, but he had lost the wider battle. American wines were anathema. He proposed a system of certification for the movement of plants, but this was rejected when, in his words, "the representatives of Switzerland and Italy, much scared at the introduction of the phylloxera, induced the Congress to sanction draconian measures against all the products of horticulture." The following year Germany, Austria-Hungary, Spain, France, Italy, Portugal and Switzerland signed the "Agreement of Berne" which set rules on notification of outbreaks, information-sharing in the search for a cure and the restriction of movement of plant materials across frontiers.*

The *américainistes* were losing the argument at home and abroad. Within the Central Commission the partisans of chemi-

---

* In spite of two major wars (France v. Prussia, Russia v. Turkey), the 1870s was a decade of international technocratic agreements. As well as those on the phylloxera, there were conventions to establish the Universal Postal Union, on international telegraphy and on global standards for the metric system. More agreements would follow on such diverse late-nineteenth-century concerns as intellectual copyright, the pitch of concert pianos, a standard railway gauge, and the suppression of white slavery.

cal warfare, its chairman Professor J.-B. Dumas, Maxime Cornu and Pierre Mouillefert, were now completely in the ascendant. American vines were unmentioned in their reports. Not only should the state use its power to restrict the dissemination of alien vines, they insisted, it should use tax revenues to subsidize insecticidal treatments.

Planchon was sidelined. He considered resignation. According to Gustave Foëx, director of the National School of Agriculture at Montpellier, he "stayed on the commission in the hope of still performing useful work, but in the face of the systematic hostility of the majority of his colleagues towards anything that might encourage the use of American vines, he realized that the only role which he could usefully play, along with his friend Gaston Bazille and some other allies, was to prevent the adoption of draconian measures against them." He could not prevent the diplomatic embarrassments of May 1878, however, when a "choice collection of American grapevines" sent by Isidor Bush of Missouri to the Universal Exposition in Paris as part of the proud U.S. horticultural exhibit was promptly impounded and burned by the authorities.*

The *sulfuristes* won. The "law of 15 July 1878" was resoundingly passed by the National Assembly and posted as a presidential decree. A month later Pierre Teisserenc de Bort, minister of agriculture and commerce, sent out a portentous circular to the prefects of France explaining the tough new moves. First was biological quarantine:

> *The new law forbids the introduction, either into the whole area or into a part of French territory, of any plants,*

---

* The rejection of all things transatlantic was total. Léo Laliman managed to get some "French wine made from grafts on American roots into the exposition by subterfuge." The jury refused to consider them.

*vine shoots, fragments of vines or compost originating in
a foreign country as well as the transport of the same
outside French territory.*

Planchon and Bazille's protestations had won a slight softening
of the line. Prefects were told: "The Minister reserves the exceptional right to authorize the introduction of foreign plants to
a precisely determined location. The purpose of this disposition
is not to put any obstacle in the way of the attempts to reconstitute vineyards in certain departments already ravaged by the
phylloxera . . ."

The legislation went further. Inspections of vineyards suspected of infection could now be made under compulsion if
necessary. The delegated authority had the power to "penetrate
properties and do the necessary work, notwithstanding attempts by ignorant or malevolent proprietors to stop them."
The results of such investigations were to be "telegraphed immediately" to Paris. Not just inspections were compulsory.
"The first duty of the government is to protect those regions
where the vineyards are yet uninfected," said the circular.
Where infection was found in a previously exempt district, "the
state can then prescribe the means of treatment which must be
applied." The government would bear the cost.

The Central Commission, up to now a prize-committee and
scientific talking-shop, became an executive agency, La
Commission Supérieure du Phylloxéra, "which," according to
Teisserenc de Bort's stern circular, "will decide, by virtue of its
office where to apply treatment, which vines should be thus
subjected to it and the preferred curative method."

The arguments raged on. Out in the fields of France it would
become a matter of sullen stand-offs between pitchfork-armed
vignerons and official eradicators. In Paris the Americanists on
the Commission and in Parliament fought to keep their alien
flag aloft. In July 1879 a move in the Senate to provide government subvention not just for insecticide but for replanting with

transatlantic vines was blocked by the minister of agriculture. The journal *L'Économiste français* commented wryly: "All the agricultural societies of the Midi have pronounced that American vines are the only way to save the vineyards of France . . . during this time the *savants* of the official phylloxera commission have refused to discuss them, they are never mentioned in their reports. In this situation we might understand the embarrassment of M. le Ministre. The legislature will hear nothing of the matter . . . but they are hardly competent to judge. There are those who cry 'Ban the American plant, there is the villain, there is the cause of our woes!' In the Midi they cry: 'American plants are marvellous! Encourage them, multiply them, create nurseries to distribute them, send ships to get more from the United States!' "

The journal, reflecting prevailing French political suspicion of all forms of state intervention, argued that it was better to let vine-growers find their own solutions – at least let them plant American vines should they choose to.

It was not to be. A second round of legislation enshrined in the "law of 2 August 1879" closed remaining legal loopholes and tightened the chemical grip. Prefects could now make compulsory inspections on their own authority where the presence of the aphid was merely suspected. The state might now also apply "administrative treatments" to phylloxera-spots in already infected departments where they represented "a danger to the public." Prefects were additionally ordered to sponsor the creation of "anti-phylloxera syndicates"* backed by state and local subsidies to treat the vines by the approved method. That meant insecticide.

The scope of the 1878–79 legislation partitioned the whole country into three zones subject to different regulations:

---

* "Syndicalization," with its socialistic undertones, carried heavy political meaning in the early years of the Third Republic. Industrial trade unions were not authorized until 1884.

A first zone, considered as phylloxera-free, within which it was absolutely forbidden to circulate French or American vines. In this zone, the authorities had powers to seek out any infected vineyards and proceed to "treatments of extinction." A second zone, more recently invaded, where traffic of alien vines was also forbidden. In this zone, the state would provide subsidies for approved insecticidal treatments.

A third zone, "completely invaded with the phylloxera," in which the importation and planting of American vines was per- mitted under licence by the Ministry of Agriculture. Insecticidal treatments would be subsidized at the request of local syndicates.

The plan looked sensible enough: a dispersed militia to lead the fight on the ground; a central body of scientists and civil servants in Paris; biological quarantine measures; state and commercial subsidies to provide weapons for the fight; conces- sions on the importation of alien vines to those whose planta- tions had already been wiped out.

There were two problems – the stubbornness, superstition and cunning of rural France; and the fact that Paris's ordained solution, insecticide, would turn out to be no solution at all. A study of the phylloxera crisis in one department demonstrates what the frock-coated *fonctionnaires* were up against.

The department of Loir-et-Cher in the mid-Loire valley is three hundred kilometers distant from the Midi. In the 1870s half of its population depended in some measure for their livelihood on tending thirty thousand hectares of vines. The resulting wine was not particularly distinguished, but was economically vital – worth thirty million francs a year. In 1875 the prefect had used his local powers to ban all imports of alien vines. Obeying the directives from Paris, a vigilance committee was formed the following year. Its composition was typical – three small landowners and two mayors. It fell to its chairman, Édouard Prillieux, to write the inevitable letter addressed to the permanent secretary of the Central Commission. It ran to a formula, like many hundreds of other dispatches from the front line. Couched in mournful *politesse*, it was like announcing a death in the family:

> *La Maléclèche, par Mondoubleau (Loir-et-Cher), 31 August 1877*
> *I regretfully announce to the Academy that I have just confirmed the presence of phylloxera at several locations in the commune of Vendôme.* Called upon by M. the Prefect of Loir-et-Cher to examine with several other proprietors afflicted vines on which the presence of phylloxera was first suspected, then denied, then asserted, I can now state with absolute certainty that the insect has already been present for several years in several vineyards.*
>
> *I hope to have the honor to next address the Academy with fuller information on the extent of this new invasion, and I hope to establish its origin, but I wished to inform you without delay of the invasion by the phylloxera of a part of France situated at the extreme limit of the vine-growing region.*

Prillieux's investigations uncovered a now familiar story – fancy varieties had been sent from Léo Laliman's Bordeaux plantation several years earlier. An anti-phylloxera committee was formed

---

* There was a scare in the Champagne far to the north-east when the Paris newspaper *La Patrie* announced the arrival of the aphid in the hallowed vineyards around Reims (it was a mistake – one of the infected communes near Vendôme, Coulommiers, was taken for the town with the same name in Seine-et-Marne). The journal *Vigneron champenois* urged all "proprietors to be on their guard against the terrible host which is about to descend on us." But the Reims Agricultural Society reported "a kind of apathy among the *viticulteurs* of the region, who seem to be asleep, as if they were guaranteed by Providence against any eventuality of the plague's arrival."

not just to passively observe and report but to fight the aphid. There was apprehensive gloom but no general alarm. Paris had prescribed the means to make chemical war on the enemy. The little Vendômois committee began the counter-attack in December 1877 with carbon bisulphide. Technology had moved on since the first crude boreholes. Patent *pal-injecteurs* had improved, and now there were also curious wheeled chariots and ploughs to get the unpleasant liquid into the soil. The treatment seemed successful enough for those who tried it. But marching up to a peasant farmer's door and demanding that he and his family embrace this expensive "remedy" was problematic. More layers of bureaucracy began to accrete – a departmental phylloxera service, then an inter-departmental one for the whole Loire valley.

The local records speak of rows and obfuscation within municipal council chambers and bruising encounters out in the fields. Members of the Loir-et-Cher General Purpose Committee, for example, at its meeting of 21 August 1878 just after the new legislation had been passed, argued that carbon bisulphide was too expensive. Why not follow the example of the Midi and replant with American vines? The prefect replied, as might be expected, that the Departmental Phylloxera Committee was opposed outright to any dealings whatsoever with *les vignes américaines* – indeed, they were why all the trouble had started in the first place. Jules Tanviray, departmental professor of agriculture for Loir-et-Cher, embarked on a lecture tour to try to get the chemical message across. It would not succeed.

The legislation of August 1879 had imparted new powers of compulsory inspection. The following month the prefect circularized all sub-prefects and mayors:

> *Inspection must be carried out especially of isolated vines*
> *– plants in greenhouses and nurseries. It has been there*
> *most often that phylloxera has shown itself for the first*

time . . . as has been demonstrated at Orléans and Tou-
louse and most recently at Dijon.

If any proprietors should refuse admission to their vig-
nobles, you must inform me immediately, and I shall refer
the case to M. the Minister of Agriculture to enable, as
per article three of the law, those measures necessary to
surmount the bad will shown by such individuals.

Bad will was becoming generalized. In 1881 the prefect had
to invoke compulsory powers to forcibly enter vineyards
suspected of infection to apply the approved "administrative
treatment." Rook-rifles were brandished at warrant-waving
eradicators. This was politically disastrous. A fix was quickly
adopted: abandon compulsion and let local anti-phylloxera
syndicates carry the burden, with public money provided to
pay for chemicals and tools (a pool of free-to-use *pal-injecteurs*
was established).

The response on the ground was totally apathetic. The first
syndicate was not formed until December 1882, in the commune of Mer on the left bank of the Loire. In the next five years
only one tenth of the Loir-et-Cher vineyard was "syndicalized."
The cultural cocktail of peasant suspicion of officialdom and
small-landowner suspicion of collectivism defeated the utopian
ideal. Growers were "inexplicably opposed to the treatment of
their vines and remained obstinately opposed to progress," the
prefect of Loir-et-Cher reported despairingly. Only with the
enemy inside their very gates would "the *vignerons* at last be
moved." By which time it would be too late anyway. But Professor Tanviray for one changed his mind. The one-time *sulfuriste*
declared in 1883 that insecticides could be no more than a
palliative. The only way out was to set up a nursery to cultivate
American vines on which local vines might be grafted. Meanwhile it was a question of keeping some vines alive by dosing
them with carbon bisulphide and waiting for the U.S. cavalry
to arrive.

Part Three

ACCEPTANCE

26

❧

The aphid's depredations in the Midi were a boon for the growers of southern Europe, naturally willing to slake the thirst of French consumers for cheap wine. By 1879 the amount of wine from Spain and Portugal being shipped into Bordeaux matched the amount being shipped out. Within three years it would be double. Marseille's docks groaned with Italian wine (and other strange things) waiting to be transported on by rail to Bercy. But just as the phylloxera ignored prefecturial embargoes in France, so it had already stepped lightly across international frontiers. The cataclysm was going Europe-wide.

The pattern was woefully familiar. An initial infection in a vineyard (or, as often, a greenhouse) to which the import of fancy American varieties a few years earlier could later be forensically traced. Slow expansion of the hateful "oil-spots" and break-out infestations invariably followed.

In Portugal, for example, the Douro valley outbreak, blithely described to the British Consul in 1872 as the result of "bad air," was traced to a fatal import at the "Quinta da Azhinheira" at Gouvinhas in 1865. A decade later the valley's uniquely valuable production had all but been wiped out. The infection was in the Dão region and reaching into the Minho.

The insect first declared its presence in Switzerland in 1871 at Baron Adolphe de Rothschild's estates at Pregny near Geneva. Maxime Cornu went to investigate. American vines had been imported from an English nursery two years earlier, he discovered. A huge swarm of winged aphids had suddenly appeared the following summer. Twelve months later the baron's vinifera and those in a neighboring vineyard had begun to wither and die. The sick were uprooted and burned with the utmost efficiency. The Swiss would adopt a compulsory eradication policy, with mandatory anti-phylloxera insurance and compensation schemes at confederation and canton level, but – as Planchon had predicted – it ultimately proved futile.

By the end of the 1870s twelve centers in Spain were blooming outwards in Catalonia and Andalucia, and multiple infections in Germany were beginning a thirty-year march. There were reports of the insect from the Crimea, Bessarabia and Romania – all of which could be pinned on the direct import of American vines or those onwardly transmitted from French or Austrian sources.

In Italy the first official report of the insect's presence came from Lecco on Lake Como in August 1879, on "smuggled" vines that had come from France three years earlier. A consultative phylloxera commission was already established, which insisted on extinction by wholesale burning or the application of insecticide. In September, Nicòla Miraglia, the director of agriculture, arrived in Lecco with a little army of exterminators who set to burning every green thing. The following spring the aphid appeared in Sicily; a few months after that in Liguria in the north-west, again to be blamed on imported vines. Its advance was slow – a pernicious infiltration rather than an invasion – but the human effects were the same: panic and denial among peasants, and demands for state action by the larger landowners.

The peasants of northern Italy would have none of it: their deep-rooted vines trellised on trees were slower to succumb than those in the Midi. In the south, however, the *fillossera* fell greedily on the vines planted close-packed on baking hillsides. By 1891 Sicily accounted for half the national infected area and Reggio Calabria for much of the rest. Not until the middle of the decade was the plague truly established in the north.

France's travails had a double effect. The shortfall in cheap wine provided a ready new market for those who could keep their vines in production. The experience of insecticides and grafting American species so painfully won on the other side of the Alps provided ready-made routes to redemption. Italy too had its battle between Americanists and chemists. An 1880 meeting of vine-growers in Turin insisted that any experiments with alien vines be conducted abroad; but the politicians in Rome were prepared to be flexible. The island penal colony of Montecristo off the coast of Tuscany was planted with 150,000 cuttings obtained from Montpellier, to be tended by convict labor. When in spring 1882, however, phylloxera was found on the roots of Taylors (an always vulnerable labrusca), the experimental prison vineyard was shut down.

In Austria the unfortunate Baron von Babo, director of the Imperial and Royal Viticulture Research Station at Klosterneuburg near Vienna, had imported American vines via an English nurseryman in 1868. Four years later his successor Herr L. Rössler woke up to find leaf-galls on his four hectares of Clinton vines. Samples had already been sent out across the empire as an anti-*oïdium* prophylactic. Splashes of infection were soon reaching out into lower Austria, and in 1875 the first outbreak was reported from Hungary, at Panscova near the border with Serbia.

The aphid's march was now spanning not just Europe, but the world.

Wine-growing in Australia had prospered since James Busby had shipped his Tarascon-nurtured vines via Kew to Sydney on a convict ship fifty years before. In 1882 the London-based *Wine and Spirit News* (an excellently inquisitive publication) dedicated a special issue to the wines of Australia. "Australian wine is year after year increasing in quantity and improving in quality," wrote the trade journal's correspondent. The reds and whites of

Victoria were "of excellent quality, although perhaps wanting in the delicacy and finesse of the genuine Burgundy or Bordeaux wines." They "remained a curiosity" however, and were near-impossible to get hold of in London, apparently because newly prosperous Australians drank all their wine themselves.

"Around 1855 a Swiss colonist came to establish himself at Geelong, about fifty miles from Melbourne and commenced to devote his special attention to vine culture," the correspondent noted. "The colonists engaged in this work today are chiefly Germans, French and Scotchmen. The system of viticulture pursued is not at present conducted on any well-defined principle, but depends rather on the fancy of the proprietor."

The aphid was already stalking the vineyards of Australia. The first infection was reported at Tyansford near Geelong in 1875. When first observed under a microscope its appearance was described as "exactly that of the oblong jellyfish seen stranded on our shores in the summer." The Victorian government passed a "Diseases in Vines Act" in 1877, but official inspectors did not turn up at Geelong for another year. Messrs. Hopton and Wallis mournfully reported that the seventy-five-acre zone of devastation was due not to "poverty or neglect" but "the agency of the winged female." It is not clear if they ever exactly saw one.* The inspectors recommended compulsory eradication at the expense of the proprietors. Improvised boxes of carbon bisulphide were also applied by a government scientist. The Geelong infection was contained – for now.

In 1880, an unofficial report into the outbreak reached a remarkable conclusion. The danger to Australia was acute. Should the infection get out of control, drought and "a sparse

---

* Genetic researchers at La Trobe University, Melbourne, declared in 2001 that DNA evidence showed that all phylloxera in Australia were parthenogenic cloned females. There were not and never had been any winged forms.

The plague struck first and hardest in south-east France. Montpellier's School of Agriculture, known as La Gaillarde from the estate around which it grew, became the intellectual citadel of the *américainistes*. The photograph shows laboratory investigations c. 1884.

PEPINIERES DE FRUITLAND.

Augusta, Georgie, le 20 Mai, 1874.

A Messieurs les Viticulteurs du Midi de l'Europe :

Lorsque j'eus l'honneur de m'adresser à vous le 20 Janvier dernier, je me trouvais sous le coup d'une désorganisation complète de la main d'œuvre. Tous les hommes valides étaient accaparés par des travaux publics considérables. Il fallut faire de grands sacrifices pour me procurer les bras nécessaires. C'est dans ces circonstances que je dus établir mes prix de revient et les baser sur une main d'œuvre dont le prix était plus que doublé.

Aujourd'hui ces travaux sont terminés, la main d'œuvre est rentreé à peu près dans les conditions normales et me permet derechef de produire du plant de Vigne à des conditions plus avantageuses.

De plus, l'ouverture d'un nouveau port de mer à proximité, la réduction considérable des frais de transport par chemin de fer par la diminution de la distance, et l'inauguration d'un service régulier de Bateaux à vapeur entre Port Royal et l'Europe, me mettent à même de livrer mes produits au Havre, à des prix considérablement réduits. Je me trouve donc heureux, Messieurs, en m'addressant à vous de nouveau, de pouvoir vous faire participer à tous ces avantages.

Je côte donc mes plants pour les fournitures de Novembre, 1874, à Mars, 1875, soigneusement embállés, étiquetés et rendus franco au Havre comme suit :

Scuppernong de premier choix, *mille francs par mille. Thomas, Tenderpulp* et autres variétés à fruits noirs frs. 1250 par mille.

Les frais de transport et accessoires étant à peu près aussi élevés pour un colis de 500 plants que pour un de mille, il doit être entendu que ces prix ne seront applicables que pour un minimum de 500 plants.

Les viticulteurs désireux de faire l'essai de ces cépages avant de planter en grand, pourraient se côtiser pour obtenir un colis de 500 plants, et dans ce cas, les plants qui le composent seront liés en bottes de 25 ou 50 et bien étiquetés, afin d'en faciliter la repartition.

Ce genre d'affaires se traite au comptant en Amérique et les prix de revient sont basés sur cette règle; je ne puis m'en départir sans m'exposer à des mécomptes, à des pertes.

Les commandes dont l'import dépasse 500 francs, doivent être accompagneés d'un mandat pour la moitié de cet import. Le destinataire me designera en même temps une maison de Banque (soit en France soit en Amérique,) laquelle honorera ma traite pour l'autre moitié, à huit jours de vue. Cette traite sera accompagnée d'un connaissement constatant l'expédition des plants, tous frais payés.

Les commandes d'un import de 500 frs. ou moins, qui ne seraient pas accompagnées d'un mandat en couvrant la totalité, seront considerées comme non avenues.

Le nouveau traité postal entre la France et les Etats Unis, permettra l'envoi de mandats sur la poste et simplifiera la remise de fonds.

Aux personnes qui désirent isolément essayer les nouveaux cépages, ou ceux plus anciens, je puis expédier de petites quantités par Messageries aux prix cotés ci dessous. Le plant sera soigneusement étiqueté et emballé pour les voyages de long cours et delivré au bureau des Messageries à Augusta, les frais de transport étant à la charge du destinataire.

N. B.—Toute commande pour ce genre d'expédition doit être accompagnée soit d'une traite sur New York, soit de bon sur la poste, soit de coupons d'obligations des Etats Unis, soit enfin de Billets de la Banque de France. Si l'envoi consiste en valeurs payables au porteur les lettres doivent être chargées.

VARIETÉS LES PLUS RÉPANDUES.

Plants de premier choix, munis de fort Chevelu.

	PAR 12	PAR 100	PAR 1,000	OBSERVATIONS.
Clinton	frs. 5.00	frs. 20.00	frs. 150.00	Derivé du type Cordifolium.
Concord	7.50	25.00	200.00	" " Labrusca.
Delaware	15.00	100.00		" " "
Diana	10.00	60.00		" " "
Hartford Prolific	7.50	25.00		" " "
Ives	7.50	25.00	200.00	" " "
Norton's Virginia	20.00	150.00		" " Œstivalis.
Salem	10.00	60.00		Hybride de Labrusca.
Warren ou Herbemont	15.00			Derivé du type Œstivalis.
Taylor	10.00			" " Riparia.
Scuppernong	12.00	80.00		} Derivés du type Rotundifolia, ne se
Thomas et Mish	15.00	100.00		reproduisent pas par boutures.
Tenderpulp	20.00	150.00		

As the phylloxera crisis deepened there was a scramble in the Midi to get hold of American wonder-vines. The entrepreneurial J.P. Berckmann, nurseryman of Augusta, Georgia, circulated this pamphlet round southern France in May 1874 pointing out how easy the new postal treaty between France and the United States made dispatch and payment. Concord vines (a variety of Vitis labrusca recommended by Professors Planchon and Riley) were on offer for two hundred francs per thousand. They turned out to be fatally vulnerable, leading to widespread disillusion and anger.

After the first wave of direct transatlantic shipments, French nurserymen began to satisfy the demand for American vines with homegrown specimens. Léo Laliman of La Touratte, Bordeaux, accused by some of having imported the aphid in the first place, was not going to let the allegations interfere with a colossal commercial opportunity.

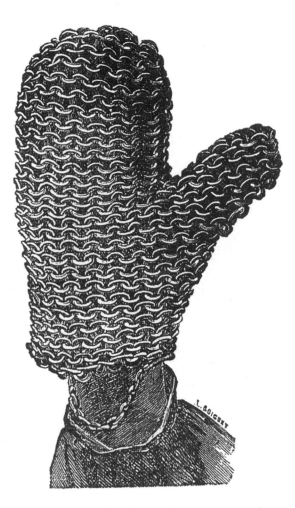

The winter-egg theory produced several proposed remedies, including this curious chain-mail glove with which vignerons were supposed to scrape growing vine-stems clean of infected bark.

population inclined to carelessness" would mean its rapid progress. It was the country's national destiny to produce wine – a destiny that could be fulfilled only by mass immigration of diligent wine-makers. Where might such skills be found? The answer was clear.

"The peasant of the Midi now pays eight sous for the wine he used to get for three, and has a fair prospect of losing his beverage altogether," the pamphlet's author, Mr. David of Melbourne, wrote perceptively. "When wine is beyond the reach of the masses, the occupation of nine *vignerons* out of ten is gone. But where shall they go? To some land of generous soil and genial climate where the phylloxera is unknown. Never will the man whose dreams have been disturbed by that hated insect run the risk of encountering him again." The Geelong outbreak must be totally extinguished, Mr. David insisted, so that ruined *vignerons* seeking a new life should sleep untroubled under the cloudless skies of Victoria.

The huddled masses of the blighted Midi preferred Algeria or Chile. But the aphid's advance in Australia was in fact very slow. It took two decades for the vineyards at Bendigo, in the Goulburn Valley north of Melbourne, and those around Rutherglen to the north-east to succumb. Uprooting and burning proved useless.* The phylloxera would keep up its antipodean advance for half a century, although South Australia remained untainted.

The phylloxera crossed the oceans to get to Australia. Just how is unknown. The deserts and Rocky Mountains that had

---

* Australia was to go through the same ritual dance as France. "Nothing now appears more ludicrous than the constantly reiterated advice tendered by authorities in vine districts free from the insect to unfortunate growers in attacked districts to persevere in their attempts at 'total extinction' and not to plant American vines 'because the phylloxera lived on them,' " wrote the directors of the Rutherglen Viticultural Station in 1901.

guarded California from the insect for millennia were breached as mysteriously around the same time. Native vines had first been cultivated in Mexico by the Spanish in the sixteenth century, and two hundred years later they had carried the "mission grape" with them northwards into Alta California. Early-nineteenth-century immigrants brought vines with them from Europe. They flourished in the frost-free valleys. It was not just a matter of California's beneficent climate and geology. The soil was aphidless.

Californian endeavors moved to a commercial scale. Vine-growing began in Napa Valley in 1838. When the United States annexed California in 1847 there were vineyards from Napa County in the north to Los Angeles in the south. The gold rush brought a thirsty stampede of incomers; most eventually slouched back to where they had come from, some stayed behind to grow vines. One of them was a political refugee from Hungary named Agoston Haraszthy who in 1854 acquired a mature vineyard in Sonoma County, run on good European lines, with a stone-built winery and cellars run in tunnels into the red, gravelly soil of the hillside. San Francisco investors bought into the business and, in a decade-long flowering before it went bust, the products of the Buena Vista Viticultural Society became famous. In 1861, on the eve of the Civil War, Haraszthy had travelled to Europe and brought back a boatload of vines – three hundred varieties, 100,000 vines altogether.

The following spring and summer a proportion of the European immigrants showed "short growth, small and colorless grapes and early yellow leaves." No one thought to dig them up and examine the roots under a microscope. It was put down to some incompatibility of the soil. More failures followed. In August 1873 the vines of a Mr. O.W. Craig on his plantation a few miles north of Sonoma Creek began to succumb. This time the Viticultural Club of Sonoma was more enquiring. Europe was in the grip of some terrible new malady of the vine. A root-sucking aphid was said to be the culprit. Charles Valentine

Riley, state entomologist for Missouri, had gone to France to see for himself, and declared the tiny insect to be the same as one endemic in the eastern United States. The roots of Craig's vines were encrusted in aphids. The Sonoma viticulturalists sent specimens to Riley for comment. They were the same. Faced with such evidence, however, California's wine-growers proved as immovable as the most obscurantist European peasant. In spite of dire warnings from scientists, a bill proposed in 1876 to "save from destruction the vineyards of California and to extinguish the Phylloxera in said vineyards" failed to pass the state senate. The collapse in French wine-making meant increased domestic demand for the Californian product in New York and Boston. No one was about to uproot and burn their vines while business was booming.

The infection meanwhile was being spread on cuttings, rooted vines, wagons and farmworkers' boots. Home-grown remedies proved costly failures. The plague, bizarrely, was blamed in agricultural journals on some fatal import from Europe: the parasite had lived benignly enough on American leaves for generations; the root form was an Old World confection, it was claimed. In Bordeaux the ever-obsessive Léo Laliman fell on such reports for his now bulging dossier of evidence. California farmers' denials continued as their vineyards shrivelled around them.

Just as in France there was the greatest reluctance to admit the presence of the shameful aphid. Professor Eugene Hilgard of the University of California College of Agriculture wrote in 1876: "The cause of this silence is twofold; First the great depression of the wine interest causing a feeling that the culture might as well be given up; second, that . . . the spread of the insect is much slower than has been the case in Europe . . . that a great deal of incredulity and wild speculation as to the cause of the dying out of the vines had become current." As in Béziers, so in Sonoma.

27

A new scientific star had joined the angels in the battle against the aphid. Pierre-Marie Alexis Millardet, professor of botany in the Faculty of Science of Bordeaux, had from 1874 onwards been quietly looking not at insects but vine-roots. The shambles of uncontrolled replanting in the Midi had painfully demonstrated that some American vines were more resistant than others.

By methodically assessing the damage – expressed in the number of root "tuberosities" – Professor Millardet was beginning to scientifically delineate the degree of resistance of each species or variety. He produced a scale from zero (vinifera) to twenty (rotundifolia). In the Vitis navy there were frail wooden frigates, lightly armored cruisers and unsinkable battleships.

The level of immunity of a variety being known exactly, it was then possible to determine to which type of soil it was best adapted. In exhaustive field trials the botanist worked out that riparia, for example, was optimized for deep and fertile ground, Jacquez for clay soils, rupestris for rocky and dry ground, solonis for wetter soil.

The reputation of riparia soared. For many *vignerons* its name, grunted in the patois of the south, would become synonymous for any American rootstock. Within two decades two thirds of American vines planted in France would be riparia. But it, and no transatlantic vine so far identified, seemed to work well in chalk soil.

There was much more to it than that. The most compatible, phylloxera-resistant American vine growing on French soil pro-

duced execrable wine. Professor Millardet had the solution –
make "completely new vines" by hybridizing them with vinifera. As he wrote in 1881: "Must it be that by giving our varieties sufficient 'American blood' to make them invulnerable to
the various plagues which their delicate constitutions cannot
withstand, some of their qualities are lost in the crossing? I am
certain they will not – our wines will retain all their principal
character." His promise seemed utterly miraculous.

When in 1878 a new affliction, le Mildiou,* began to show
itself in France, Millardet went further, seeking to breed resistance to the new parasite in a hybrid vine, as well as to that
old fungal enemy the oïdium. Working from 1880 onwards
with M. le Marquis Charles de Grasset, an entrepreneurial
wine-grower of Pézenas in the Hérault, Millardet produced
eight hundred interspecific crossings. More eager hybridizers
got into their greenhouse-laboratories and experimental plots
– the lawyer Victor Ganzin, owner of vineyards in the Var
who had successfully produced an Aramon and rupestris cross
in 1877; the engineer Georges Couderc, viticulteur of the
Ardèche; and the Swiss-born professor Gustave Foëx.

It would seem for a time that these miracle hybrids – so-
called producteurs directes – might indeed repel the underground enemy while delivering wine as delicious as any from
the noblest European variety. They would not. Those old bug-

---

* "Downy mildew," caused by a fungus, Plasmopara viticola, related to the
potato late blight pathogen, which turned out to be, after oïdium and phylloxera,
yet another disastrous biological import from America. In 1878 the undersides
of vine leaves were first observed covered in a grey-white down. In warm, wet
weather whole grape clusters were consumed by it. After a chance observation
by Millardet by a footpath in a Médoc vineyard, Château Ducru-Beaucaillou, of
strangely unafflicted vines over which lime had been dispensed from copper cans
(evidently to deter "grape thieves"), it appeared that application of copper sulphate to the foliage destroyed the fungus. Downy mildew meanwhile spread all
over France and western Europe. In 1885 Millardet patented his famous bouillie
bordelaise (Bordeaux mixture), a compound of copper sulphate, slaked lime, salt
and water as an effective cure.

bears, the undertastes of "fox" and black currant, kept coming through. Direct producers would have an undistinguished career, although they had their champions for many years. The work of the hybridizers however would prove vital in providing rootstock that would bear European scions and prove adaptable to the soil and climate of France, Europe and eventually almost the whole world. The future of vinifera, the future of wine, depended on it.

As Professor Millardet published his first theoretical findings, the future of wine-making in southern France depended on whatever could be made to grow. In spite of the horrors of the tasting of American wines at the Montpellier Congrès Viticole of 1874, four years later millions of American vines had been and were continuing to be planted in the soil of the Midi.

"The phylloxera crisis has just entered a new phase," *Le Temps* told Parisian readers in August 1878. "The Hérault, the *département producteur par excellence*, has decidedly thrown *le manche après la cognée* [given up entirely], or so it has declared through the pen of one of its most eminent *viticulteurs*. Gaston Bazille having lost all hope of checking the plague's advance with insecticide, now says he has renounced the struggle and put all his faith in American vines . . ." He would do so with all the zeal of a convert – he planted Jacquez and Herbemont, thousands of young vines went into his domain at Saint-Saveur near Montpellier. In three years' time they would bear grapes. The sternest test of all would come when the *négociant* arrived to make an offer for the wine.

The loudest propagandist for the transatlantic immigrants was a *viticulteuse*, the Duchesse de Fitz-James, owner of a huge domain at Saint-Bénézet near Saint-Gilles in the Gard. There, from 1872 onwards, she had been creating one of the biggest plantations of American vines in south-east France.

Mme. la Duchesse had arrived in Paris as Marguerite Augusta Marie Löwenhjelm, daughter of the Swedish minister to the newly imperial court. In 1851, aged twenty, she had married Édouard Antoine Sidoine de Fitz-James, a descendant of James, Duke of Berwick, bastard son of the exiled King James II. The "intelligent and fearlessly innovative" Swede with the Jacobite name was an ornament of Paris society. She wrote books on equitation and horticulture and each autumn decamped, as was the custom, to the family domain in the south. There seems however to have been an estrangement from the duke. She grew vines, naturally, and her vinifera had been in the front line of the aphid's first eruption. The duchess had embraced insecticides almost from the first days of their promotion. The first crudely regulated doses of carbon bisulphide killed both the aphid and her vines, until in 1874 a local chemist, a certain M. Fichet, prescribed his own "secret liquid" that seemed to hold the line – although at considerable cost. Alien vines for now remained unthinkable.

One morning an itinerant salesman of American stocks had ventured up the long twisting road to the duchess's house in the hills from which she rode each morning to supervise her vineyards on the hot plain. She had gruffly dismissed, as she wrote, "the first merchant of plants who presented himself – who must remember how he was treated, not even being allowed to leave his name and address." But the scale of chemical warfare was too overwhelming, the battle too intense, skilled labor to apply the horrid liquid impossible to find, the generalship in the fight too demanding. "What if one should fall ill?" she wrote. "One's crop was at the mercy of a moment's forgetfulness, or of some crisis such as a war or a death in the family."

The flamboyant duchess became the "Americanizer" (her word) in chief. Her powers of persuasion were formidable, her eye for detail hawklike. In 1876 she built a narrow-gauge railway (*un Decauville*) to convey visitors round her remade vineyards. Every railway station in wine-growing regions should have five American plants in a little plot, she proposed, so that travelling *vignerons* could see how they flourished alongside the morbid vinifera. The idea was taken up.

As for insecticide, it might work for the "*grands crus* and small operations delivering big revenues," proclaimed the Duchess, "but only in American vines will producers of *vins ordinaires* find salvation." Those who had already been converted pronounced the bottled results to be delicious.

The Paris press was hugely skeptical. "We continue to be mistrustful not only because of the incessant contradictions of the partisans of these varieties," declared the country life correspondent of *Le Temps* in August 1878,

> but especially because of the immodest eulogies which surround their products.
>
> On paper they are declared to be as good as anything else – but when one comes to try them it's a different matter. We have participated in at least half a dozen tastings of American wines. My own opinion counts for nothing, but I can tell you that not one of those who took part had the courage to empty his glass.
>
> It may be right to admit that this red liquor, let's not prostitute the good name of wine by calling it so, has a certain alcoholic strength, as much as fourteen degrees. This is enough apparently in the eyes of the viticulteurs of the Hérault to justify their decisions that have resulted in such a terrible wreck.

The chemists were winning the taste war.

A week later, on 26 August 1878, Jules Lichtenstein made a counterblast in the same rustic column:

> You forget in condemning this red liquor which you refuse to call wine that ten million American vines have already been planted in the department of l'Hérault alone, of which

*a third at least are destined to serve as simple rootstocks.
While the root may change, the grape remains the same.
This reconstitution continues at a great pace. M. le Sen-
ateur Pagézy has thirty hectares – and will this year make
a harvest of French wines grown on American roots . . .
Next month from 4 to 6 September there will be a wine
congress in Montpellier and an exhibition, not only of
American grapes but of* French grapes come from grafts
on American vines *[Lichtenstein's emphasis]. Let the
doubters of Paris come and see them – we are not that
difficult to get to – and be convinced that a Pinot remains
a Pinot, a Cabernet a Cabernet.*

*The vines of France are doomed. I have become con-
vinced of this during ten years of observations and research,
but the wines of France will live again, reborn on the resistant rootstocks of America. We will surely pass through a
time of testing when our Bordeaux and our Burgundy may
be slightly inferior because of the youth of the plant, but that
is a defect that time will correct year after year . . .*

The Parisian newspaper could not agree. The "fight must continue [with insecticide] and any recourse to America forsworn
until, should it come, the very day of defeat," it editorialized
in return. The spread of American vines had just served further
to spread the disease. The claims for grafting remained unproven. Where was this untainted wine raised on foreign roots?
Had anybody drunk it?

There was some hope. "The work of M. Millardet seems to
be taking research for a defense in another direction," said *Le
Temps*, "and we cannot entirely despair that the offer of the
prize so far unawarded will not one day result in a practical
and economic remedy that will save our French vines entire
from the tip of their stems to the base of their roots . . ."

* * *

Jules Lichtenstein's invitation to snobbish Parisians to descend on Montpellier was not rewarded by a stampede south. The Congress of September 1878 was the by-now-standard party rally of the Americanists. The price of vines was falling, cuttings were coming on a large scale from local nurseries, the disasters and misapprehensions of the past were behind them, Gaston Bazille told delegates. "Your vine is dead. Dig it up!" he declared – and if there were doubts about wine from American grapes, there was already an answer. "Put into effect the excellent advice you have heard in the past few days," Pagézy declared, "order grafted plants now – and at a cost of between four to six hundred francs a hectare you will have reconstituted a vineyard where you will cultivate vines just as they were before the phylloxera came!"

But where was the proof? Juicy bunches of table grapes from grafted vines were on display at the conference, but not their fermented product. There would be no public repeat of the *dégustation* disaster of four years before. The elderly but still pioneering former senator Pagézy was about to harvest the fruit of French vines. "Aramon and Cariganes grafted on American roots," Bazille told the congress. He would make his wine that autumn. "There will certainly not be enough to fill anyone's cellar, but the first and the most difficult step will have been made." The strange new vintage would not be ready for tasting quite yet.

28

The aphid cared nothing for the official maps of the *départe-ments indemnés* and the grey-tinted *zones phylloxérées* now glumly pinned to the walls of mayoral parlors across France. Nor did many increasingly desperate vine-growers. Smuggling was rife, the system of certification for the movement of alien vines had descended into bureaucratic chaos. A glance at *la carte du phylloxéra* diligently updated by the Superior Commission told its own story. After more than ten years, the initial blotches of infection in the Midi and the south-west had bloomed ever out-wards to join up around Toulouse. By 1880–81 the pest reigned in southern France from the Alps to the Pyrénées. It was moving north-west and north-east in two huge pincers, indifferent to the largely vineless uplands of the *massif central* in the middle. The vineyards of Burgundy and the Médoc would be next.

Planchon had perceptively written in 1876:

> *Burgundy hopes to defend itself by snuffing out the first signs of the evil, even at a very high price. It quakes at the idea that American vines should encroach on the domain of its vintage wines, even as auxiliaries in the fight. The Bordelais, just as proud, but already profoundly affected, call on the help of exotic vines as a last resort, but with the secret hope that the destruction of the winter egg will save its native vineyards.*

The aphid's advance in the Médoc had been comparatively slow, but by 1879 it was at the gates of Margaux and Pauillac.

The grandest of all France's vineyards would have to make a stand. There were ample cash reserves on deposit from the good years. The defense would be chemical. An Association Médocaine Contre le Phylloxéra had been established in 1876 on the government prescription. Three years later, a number of medium-sized proprietors formed syndicates to buy carbon bisulphide and injectors. The grandees were suspicious of such socialistic confederations. If the pest came they would fight it with their own resources. As the correspondent of *Wine and Spirit News* informed concerned English consumers:

> *The proprietors in the Médoc have by no means lost their confidence. From all that we could learn, this question of the phylloxera at the present moment is in much the same state as that of the* oïdium *was . . . then we heard on every side the most gloomy predictions and anticipations of the probable total annihilation of the vine.*
>
> *Soon after a very simple remedy was discovered. Perhaps it may appear a little bold to say so but we are disposed to think that the question of the phylloxera has passed into the same stage and that by the regular application of such remedies as sulphur of carbon it may be possible to keep under, if not entirely destroy this insect.*

Representatives from Latour and Mouton had meanwhile already been to Cap Pinède, the citadel of sulphurization, to consult Professor Marion. The defense was ready when the aphid at last arrived in the Haut-Médoc in July 1879 amid general uproar. M. Lietaud, *moniteur-en-chef* of the P-L-M railway's *service du phylloxéra*, arrived at Pauillac a week later to personally direct the injector-wielding labor gangs. The first application of carbon bisulphide went into the famous quartz-pebble soil of Château Latour's vineyards in November 1880, followed by a second in May 1881, a third the following August. It was getting expensive.

The Comte de Courtivon, Latour's director, wrote portentously at the outset of the fight: "If we are going to defend ourselves, if we are going to survive at all, it will be necessary to perform this operation every year. Without exaggeration this is going to cost more than five hundred francs a hectare. The battle is joined. We must sustain the struggle without prejudice or illusion, but also without despair . . ."

The results were not encouraging. The Bordeaux correspondent of *Wine and Spirit News* gave London readers a gloomy prognosis: "The vines so treated are never completely freed from the insects. New colonies spring up in the summer . . . and the least practical eye can see the swarms of insects that infect the vineyards. This reinvasion is the result of a few phylloxera that are left behind on the roots by a treatment which must of necessity be imperfect." Applying the more effective insecticide potassium sulphocarbonate was even more expensive. Latour later invested in its own steam engines, pumps and pipework with a team of engineers to run them. Others like Mouton hired entrepreneurial contractors to get the chemical into the ground.

The humblest *vignerons* meanwhile, working their small but high-value patches of soil, did nothing, suspicious of insecticide, suspicious of the government, suspicious of the big owners, suspicious of the Americanists, suspicious of each other. The rate of "syndicalization" was very low. Their vines died.

Burgundy too had seemed exquisitely immune. But as the pest's *avant-garde* crept up the flanks of the Rhône, past Lyon into the valley of the Saône, on towards Villefranche and Mâcon, surely even the most blinkered could see the enemy was at his gates. In the Beaujolais the first spot of infection was reported at Villié-Morgon in summer 1874 – the discovery that had so discomfited Victor Pulliat before his appearance at the Montpellier Wine Congress a few weeks later. The pace of the aphid's advance was slow but inexorable. In the south, the fêted vines of Brouilly, growing on their famous clay-covered dome of granite, were almost all obliterated by 1879.

In this southern outpost of Burgundy, as everywhere else when the phylloxera first began its underground attack, witch-doctor remedies were invoked. Snail slime was hailed as the sovereign remedy. In the Beaujolais schoolboys were conducted twice daily from their classes to urinate over vines (to observe niceties, perhaps, only urine of the "masculine sex" was judged effective). Huge volumes of horse urine were retrieved from the barracks of the 11th Cuirassiers at La Part-Dieu in the northern suburbs of Lyon to be sluiced over unfortunate Gamays.

Under the legislation of 1878–79 the fight became more me-thodical. With its one-hundred-franc-a-hectare state-funded subsidy, carbon bisulphide unsurprisingly was the chosen weapon. The defense syndicates with their injectors and sul-phurous ploughs held as firm as they might. American vines were regarded with disdain, their champions in the Midi de-rided as "ruffians" with Victor Pulliat as their treacherous ally.

To the north the same slow-motion process of denial, anger and ultimate acceptance of a sentence of biological death was about to be played out in the low, rolling hills of the Côte d'Or. The Burgundians' defenses had long been put in place. A departmental Vigilance and Study Commission had been formed in 1874. The following year a prefecturial decree had banned the import of alien vines. The Burgundians seemed to be admirably open-minded. Officials had already visited both Montpellier to inspect the Agriculture College's *vignes d'expér-ience* and, courtesy of the P-L-M, Marseille to see Professor Marion's carbon bisulphide-dosed plots at Cap Pinède. Know-your-enemy recognition charts depicting the aphid in all its life-stages had been distributed. The wait for the devastator's arrival was not overlong.

The first infection was found in July 1878 by a M. Viard in his vineyard at Mersault. The prefecturial response was urgent: a squad of six red-trousered infantrymen and a sergeant were sent from the garrison at Dijon to "seal off all access to the infected vine." P-L-M technicians arrived soon afterwards with their sinister-looking injectors and barrels of chemical. In their goggles and rubber aprons they might have been men from Mars. When new *taches* were found nearby, local laborers were reluctant to work with the insecticide. The military were ordered to perform the statutory "administrative treatments." Clumsily they did so. After a month it was announced that "all plague-spots had been destroyed."

They had not been, of course – within a year the aphid was in the sacred vineyards around Beaune, where once again insecticide was to be sloshed around as the only answer. There was bitter resistance. At Savigny-lès-Beaune for example, in summer 1879, the arrival of Departmental Inspector M. Bonelli with a squad of "foreign" gendarmes (they had been brought north from the Midi) was greeted by a violent uprising. The *brusquerie* of Bonelli and his alien flying squad became extremely unpopular. M. Duval, the harassed prefect, had to suspend operations until "after the grape-harvest, when it was to be hoped the *vignerons* would be more amenable." He must have come close to despair when the growers of Pommard declared the "phylloxera was the friend of the vine – since its arrival the wine had actually got better."

Forging a common front proved as problematic as it had anywhere else in France. Communal vigilance committees spent their time philosophizing like the revolutionary cadres of a guerrilla army. There were many conspiracy theories. The phylloxera had been introduced by the "reactionaries" after the republican parliamentary triumph of 1877. According to the Bonapartist journal *Bien public* of Dijon, inspections and compulsory treatments were a plot by bourgeois proprietors to dispossess the humble and employ them as common laborers. State intervention was a mask devised to help increase the profits of the *rentiers*, the chemical manufacturers, the railway companies, fraudulent nurserymen, the sharp-faced middle-men of

Beaune, Bercy and Bordeaux. The lawyers moved in, arguing that compulsory searches were legal only in as yet uninfected districts. The law of August 1879 shut this loophole, giving prefects new powers to search and destroy.

At Bouze-les-Beaune there were violent demonstrations. A "Bonapartist agent" was accused of ramping up the trouble between the "Mayor's party," armed with all the inquisitorial impedimenta of atheistical science, and the much more numerous opponents of chemical treatment. The gendarmerie were called out to restore order. At nearby Puligny meanwhile "the inhabitants accepted it without a murmur, the chiefs of the party being partisans of treatment."

At Chenôve south of Dijon on 18 August 1879 a crowd of 160 *vignerons* chased the sulphurizers out of the commune. "Among the demonstrators, there were those who said we were the villains, more to be feared than the phylloxera," according to the police report. The rising's leader, a small proprietor, was arrested. When acquitted by a sympathetic jury he was carried through the streets in triumph. It was, said the Dijon historian Robert Laurent, the "administration's final débâcle." The rumbling land war was just too dangerous for compulsion to continue. Bonelli was dismissed. (He was replaced by a Parisian – no Burgundian evidently would act as phylloxera-finder general.) The soldiers were confined to barracks.

The fight was passed to free-standing anti-phylloxera syndicates with a subsidy for insecticide of eighty francs per hectare. They began with admirable solidarity, but little by little began to lose heart. The richer growers could afford to defend their high-value plantations, hiring labor gangs to do the unpleasant work if necessary. The lowlier meanwhile, "always defiant and incredulous," in the words of the Vigilance Commission's *rapporteur*, "still refused to take part." For the grandees of Beaune the idea of planting American vines remained an abomination.

* * *

As they remained for the Superior Commission. "Science will emerge victorious from this great struggle," read its report to the agriculture minister for 1881.

> On every side confidence is returning to the vine-growers of France, so wise, so patient, so industrious. There may be differences of opinion on where to find salvation – but all are united in their certainty that happier times are coming . . .
>
> From everywhere comes evidence of satisfaction with the happy results of [chemical] treatment. Where once there were conflicts between the authorities and those who refused to let their vines be treated, today vine-growers are begging our agents to come and defend their vines . . .

Not once did the report mention redemption from America. Everything in the sulphurated garden was rosy.

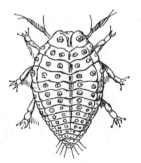

29

A showdown was coming – it would be very gentlemanly, but it would have to come nevertheless. More than ten years since the devastator's first appearance in France, the officially prescribed defense was failing. The quarantine decrees had hobbled the *américainistes* and directed state subsidies towards anti-phylloxera syndicates, the sulphur merchants and their railway company allies. The aphid's advance continued.

Ruined proprietors were now more than ready to disobey government directives. Instead of armed clashes, however, the battle of ideas would now be played out in a series of set-piece conferences, the first of which was held in Lyon in September 1880. The animals of M. Pezon's travelling menagerie were confined to their cages. Brutus the lion and Flambeau the tiger would not be making their matinée performance today. In September 1880 the Viticultural Society of Lyon had hired the famous showman's tent in which to hold four days of debate on the only issue of the day. It was packed, the participants animated. The Lyonnais had seen two thirds of their vineyards destroyed. A petition from "twenty thousand heads of families" had already been sent to the Ministry of Agriculture in Paris begging for financial relief – "otherwise nothing would stop the emigration of *vignerons* ruined by the ravages of the phylloxera."

Chemicals were too expensive, claimed a procession of delegates. They were too difficult to apply. They were a "sword" in the fight, but resistant vines were the only true shield. All the *américainiste* stars were on parade to reinforce the message – Planchon, Pulliat, Bazille, Lichtenstein, Foëx, Vialla, even the increasingly eccentric Léo Laliman. Depositions were made on the merits of both direct producers and especially of grafted vines. The ban on Americans must be lifted, delegates resoundingly concluded, and a "little nursery of mother vines established destined to provide subjects for grafting."

The Lyon congress ended in chaos. At its very last session Prosper de Lafitte, president of the Lot-et-Garonne Vigilance Committee, rose to speak. His peroration was doom-laden: insecticides were merely a palliative, they must ultimately fail. American vines were a leap into the unknown. He dwelled interminably on the mysteries of the winter egg as the only hope of salvation. Outside, a new audience had gathered, waiting for M. Pezon to reclaim his tent and start the evening performance. The showman released his monkeys, which swarmed round the orator, whooping and grimacing. "M. de Lafitte continued to speak," according to the Duchesse de Fitz-James, "firm and noble to the last."

The exotic vine lobby meanwhile was attracting its own share of eccentrics. M.A. Lavallée, the secretary of the French Horticultural Society, had managed to obtain samples of vines from "northern China, Manchuria, Korea and Japan" in the early stages of the aphid's northward march as a potential line of defense. In 1875 he had dispatched some to Mme. Ponsot in the Libournais for trial. When after three years her Asian vines were found to be as yet free from phylloxera, there was great excitement. In 1878 a "*viticultor of the Hérault*" (he preferred to remain anonymous) asked perfectly reasonably in the *Journal d'agriculture pratique*: "But what sort of wine do they make? Surely we should commission our consular agents in these countries to get hold of some samples. Otherwise will we not end up with some hideous brew, as we have already done with vines from the new world . . . and crown our vineyards with another *popote sauvage* [savage headdress]?"

In 1879 the African explorer Thomas Lécart published news of his discovery of several mysterious species of vine that grew profusely in Sudan. Professor Planchon was skeptical. "In spite of the respect due to the memory of a victim of scientific experiments [Lécart had subsequently perished in pursuit of further explorations], these African vines, if they were vines at all, could never be cultivated in a European climate," he warned. "Neither Bordeaux, Burgundy nor the Champagne" should put their faith in these "enigmatic plants."

There was hope in India, it seemed. The *Wine Trade Review* reported on 15 September 1881:

> *The Maharajah of Kashmir is prosecuting his grape grow-ing and wine-making projects with much vigour. Mr G. Ermers, superintendent of HM's agricultural works, re-turned from France by the last mail with two practical vintage-tasters.*
>
> *As the growth of grapes is expedited, Mr Ermers fully anticipates making some excellent wine. These are from French vines planted in Kashmir two or three years ago. Mr Ermers took with him to France a large number of cuttings and seeds of Kashmir vines which the French growers hope will help them to combat the destructive phylloxera by grafting French vines upon Kashmir stems.*

In the late summer of 1881 chemists and Americanists squared up for a climactic clash at the grand International Phylloxera Conference, due to be held at Bordeaux that autumn. This time the attentions of the metropolis and much of the wine-drinking world would be riveted by the proceedings.

Two delegates left lively accounts of those blustery six days. The Duchesse de Fitz-James filled half a book with her robust narrative. Roland Trimen, entomologist of Cape

Town,* was also there to make a report – seconded by the government of Cape Colony, South Africa, as its special commissioner. Trimen seemed quite impressed: "Notwithstanding the remoteness and general discomfort of the dreary Salle de l'Alhambra where the meetings were held, and the almost unbroken wet weather," he wrote, "the whole affair [had] an earnestness and reality too often absent from gatherings of this nature."

More than seven hundred delegates, including representatives from Spain, Portugal, Austria, Russia and the United States, milled through the vine-bedecked halls of the Salle de l'Alhambra on the rue Judaïque, otherwise a palace of varieties. William Thistleton-Dyer, assistant director of Kew Gardens, represented the governments of the Australian colonies. There was one topic of conversation, according to Trimen: "the phylloxera, and the phylloxera only." When he told his French hosts there were no aphid devastators in South Africa, they lost all interest in anything he might have to say.

Specimens on show illustrating the biology of phylloxera and its method of attack were "especially deficient," Trimen thought. Scant effort had been made to prepare these clumsy tableaux, probably because they were by now "too painfully familiar." There was a display of chemicals "recommended as being effectual," plus the various injectors, chariots and

---

* Famous as the discoverer of the elusive Brenton Blue butterfly, Roland Trimen was also a lively correspondent of Charles Darwin on the matter of the pollination by insects of exotic South African orchids. Like those in Melbourne, the Cape Town authorities saw an opportunity in the phylloxera crisis to "induce small wine-farmers from Southern France and Italy" to emigrate to the colony and thus boost wine production – the "most remunerative of all colonial agricultural pursuits." "Immigrants from Holland and England are of little use knowing nothing whatsoever of the work to be done in vineyard and wine-store," wrote an official in November 1881. Phylloxera reached South Africa five years later.

animal-drawn "sulphurous ploughs" devised to apply them.*
"Little attempt had been made to explain their properties or
effects," the entomologist noted.

Perhaps the *sulfuristes* with all their government and commercial firepower thought they need not try too hard. By contrast "the most instructive portion of the exhibits was a fine
series of French-grown American vines both as direct-yielders
and as stocks for grafts from European varieties and of the
various appliances invented to expedite the process of grafting," Trimen recorded.

The congress's list of exhibitors indeed counted numerous
small proprietors from the Midi and Gironde proudly display-
ing American vines and samples of their fermented products.
The Montpellier School of Agriculture put on a fine display of
grafting tools and novel techniques. Messrs. Saintpierre and
Foëx presided over tempting samples of American grapes, but
evidently chose to keep any wine made from them under the
counter.

Two exhibitors however, M. Molinier of Château de Pignan
near Montpellier and M. Julian of Villeneuve-lès-Maguelonne
in the Hérault, had brought "French wine coming from grafts
on American plants." A tasting panel would judge them later
in the proceedings. Its pronouncements would surely be the
decisive moment of the conference. Would they taste of
"fox"?

Armand Lalande, newly elected republican deputy for the
Gironde, declared the congress open on the morning of Mon-
day, 10 October. Cardinal Donnet, archbishop of Bordeaux,
and General Dumont, the garrison commander, beamed from
the rostrum, flanked by a pantheon of other Bordelais digni-

---

* The proclaimed merits at the congress of the "Salvator Vitis" chemical plough
invented at Floirac in the Gironde included its ability to wipe out that other
"vermin," the field-mouse.

taries. Pierre-Émanuel Tirard, the minister of agriculture, was indisposed.

In his opening address Lalande lavished praise on those who had "led the gigantic fight against this microscopic enemy." The Duchesse de Fitz-James received special mention, amidst loud applause for her noble part in the struggle against the pest "which had inflicted enormous losses on French viticulture estimated at more than five billion francs." Delegates' cheers turned to groans. Someone suddenly jumped up from his seat shouting, "If you will let me speak for just ten minutes, the phylloxera will cease to exist! Death without mercy to the *aphidien!*" Lalande was unmoved – the schedule of speakers could not be interrupted. The mystery remedy went unannounced. "The phylloxera lives on, at least until tomorrow," noted the reporter from *Le Temps*.

Four days of formal presentations, speeches and argument began, followed by two days reserved for inspection tours of blighted domains in the Médoc and Saint-Émilion (spent huddling in wine-cellars, according to Roland Trimen – the weather was too inclement to search for autumnal insects). There were discussions on submersion and the culture of vines in sand – all very worthy, but according to the delegate from Cape Town it was day three that brought real fireworks:

> *The extremely animated discussion which ensued – especially on the report of the commission on the American vines – showed very plainly that the French wine growers may be grouped in two grand divisions, viz: (one) Those who still advocate the freeing of the European vine from its enemy by the application of insecticides, and (two) Those who are strongly in favour of the more recent plan of restoring the vineyards by substituting American for European stocks.*

The rival camps had spent the weeks before combing the Gironde and the Midi for partisans willing to declaim the efficacy of their own remedy. First it was the turn of the chemists. The benefits of carbon bisulphide were rolled out in a parade of testimony by Girondist *vignerons* – year one, vines began to recover; year two, the roots were re-formed; year three, production of grapes was "almost normal." There was no other path to redemption, argued the chemical's apostles. M. Chenu-Lafitte of Bourgeais had struggled with American Jacquez vines for six years, but they had brought him only disaster, "wiped out," he said, "every summer by mildew." The *américainistes* were fraudsters, one chemical-partisan alleged; they were mere "wood merchants" trying to make a killing out of a giant "South Sea bubble" of speculative fever.

The mirthless chemists were well organized, disciplined, always in attendance, expressing approval or disapproval as a block, according to the Duchesse de Fitz-James. Their position in the hall made her think of a *battaillon carré* (the defensive infantry square of the Napoleonic battlefield). A disappointed former Americanist, M. Laroque of the Hérault, now an advocate of sulphur and stern opponent of alien vines, came in for a special pasting from the Swedish aristocrat: "That *hussard de la mort* [pall-bearer], fighting the phylloxera on the dead body of the victim itself," as the duchess described him. "If he is so right, why had the masses not rallied to his banner? . . . However explosive and inflammable the hidden thunder called carbon bisulphide, it has blown up neither the Americanists nor the phylloxera. But it is said (and there are people mischievous enough to say so) to have killed two hectares of a famous château."

The Americanists by contrast were "happy and confident," sitting where they liked, coming and going as they pleased, and seemed casually confident of final victory. All the stars were there. When their turn came on the third day, "Montpellier's brilliant orators," as the duchess swooningly described them, let loose with the now familiar message. American vines resisted the aphid,* and there was nothing wrong with wine made from their grapes.

Gaston Bazille gave delegates a seductive description of the autumnal countryside he had left behind to attend the conference. Three weeks before, the dusty roads of lower Languedoc had been clogged with *les pastières*, the traditional carts used for bringing in the harvest. They had been loaded with American Jacquez grapes, enormous quantities of them, so many that "the harvesters themselves had marvelled at their abundance."

The grapes had been pressed – the moment of décuvage *had come. The wine merchant was at the door, a most unsentimental personality. "Beautiful color," he said, "a little harsh perhaps, lots of alcohol, a good* vin de coupage† *– what price are you asking?"*

The proprietor hesitated. He's selling wine from his Aramon at twenty-five francs the hectolitre. He's too embarrassed to set a price. "Fifty francs," he said at last, astonished at his own boldness. His neighbor, seeing the price accepted so readily, asks for fifty-five francs, his neighbor asks for sixty. I am assured that right now it is selling for seventy francs.

"What more do you want?" concluded Bazille. "Wagons full of wine worth seventy francs a hectolitre! The cause is won! What need is there to speak any more of sulphur?"

---

* Roland Trimen gave an unabashedly Darwinian explanation for this in his report to the governor of Cape Colony: "The immunity of American forms is simply due to their having been obliged for innumerable generations to resist the insect's assaults in their native woods." He thought it "indisputable . . . that the French viticultor should thankfully accept the happy result."

† Wine for diluting. J.-A. Barral, vine-grower of Faugio in the Hérault, recalled a meeting with a wine merchant in 1880 who examined his wine made with Jacquez grapes. " 'What rich wine,' he said. 'It is detestable,' I replied, 'but it comes in six colors.' He readily understood that by adding water and a little alcohol one could make six times more trade wine."

There was a third group between the Americanists and the chemists whose special character, according to the duchess, was "the denial of all hope." Its chief was Prosper de Lafitte, winter-egg fanatic and victim of M. Pezon's monkeys. His gloomy followers reminded the duchess of "the thousand-headed hydra – cut one off and another one grows."

"These *désolés* should make a voyage through the Hérault, Gard and the Var to see the giant strides American vines had made," the duchess declared. "Come to Pignan and Saint-Bénézet where they would see Aramon on riparia, Oeillades on Taylor – all flourishing . . . A systematic incredulity would be required not to believe on seeing these beautiful vineyards that we are at the Holy Saturday of viticulture, on the very eve of its resurrection."

It was Saturday, 15 October when the congress reached its final session. Due to an "excess of eloquence" proceedings had overrun by two days. It was time to vote on its conclusions and recommendations. "One of these was firmly resisted," said the correspondent of *L'Agriculture pratique*; "it required the chairman's intervention to get it admitted." It was to move the government subsidy from chemicals to American vines. The motion was passed with a thumping majority.

According to the duchess, "many Americanists who had come far had been obliged to return to their homes . . . At the very hour that their success hung in the balance at Bordeaux they had gone to draw from their barrels or sell this American wine whose quality, even the very existence of which, was being fiercely questioned."

Indeed it was. Where was the proof that a future mortgaged on American roots would produce anything fit to drink? There were several "grafted" wines on display, but in spite of furtive sippings and swillings in corners of the Salle de l'Alhambra, so far no one had been allowed to formally judge them. Delegates heard a presentation by M. Bisseuil from the Charente-Inférièure on behalf of the learned M. Xambeu, professor of chemistry at the college of Saintes – who had somehow "been prevented from attending the *séance*."

It was a soulless analysis of the sugar and alcohol content of 1880–81 wines from American vines growing in the Cognac region – including several produced from vinifera grafts – appended to which was a clinically terse judgement of their quality:

> *Observation 1: On la Folle-Blanche* [the famous Cognac grape] *grafted to Riparia: The grape has lost none of its quality.*
>
> *Observation 2: On Quercy, Aramon, Cabernet grafted onto American varieties, the grape is good without any appreciable modification.*
>
> *Conclusion 1: Grafted Folle-Blanche gives a wine identical to that which is otherwise produced in the Charente and which make the best* eaux-de-vie.
>
> *Conclusion 2: Quercy, Aramon . . . grafted on American vines give wines comparable to the red wines of Saintonge.*

That was the absent professor's opinion. What would P.-A. Labrune, "president of the jury on American wines and on French wines on American roots," have to say? "These tastings have been the most interesting I and my colleagues had ever undertaken," he told the expectant assembly. Herbemont and especially Jacquez direct producers made something "comparable to the *petits vins* of the Gironde," he announced. What about grafts? The moment of resurrection promised by the Duchesse de Fitz-James seemed to have come. "As for French wines produced by vines grafted onto American roots, they seemed to present their original character," said Labrune.

And that was it. No exquisite deliberations on taste or quality, no judgements on *saveur* or *arôme*. With "infinite regret" the jury had decided it could not judge these wines formally, because "they had been made in too small quantities, drawn too soon in order to be ready for the Congress, fermented too soon as well as being altered by the journey," said Labrune.

The jury did however give a gold medal to the Girondin pioneer M. Piola of Saint-Émilion, for his "exhibition of such a wide range of wines."

But Roland Trimen and William Thistleton-Dyer were able to pass their own judgement – on wines made directly from American grapes at least. Purply, richly alcoholic draughts of Jacquez and Herbemont were proffered to the colonial visitors, "wines which it is anticipated will replace to a considerable extent the coarser descriptions of *vin ordinaire* and *vin du Midi*," said the butterfly expert. "We agreed that both were extremely harsh and unpalatable – suggesting the mixed flavors of vinegar and mulberry juice – although of the two the Herbemont was preferable."

The judgement of his colleague from Kew was no less severe: "To our taste these wines were far from palatable, being exceedingly acid, and with a peculiar mawkish flavour," said Thistleton-Dyer. "Doubtless, however, wine of this rough type, however distasteful to a cultivated palate, will not be unacceptable in a country where the poorest peasants and even their children use wine as their habitual drink," Trimen further informed his sponsors in Cape Town, "and it will at any rate be infinitely preferable to the watered and adulterated fluids with which it has of late become the practice to supply the great and unceasing demand for ordinary wine."

He was absolutely right. As cheap wine dried up, a strange new industry began to bloom across France. In farmyard kitchens and backstreet factories, wine was cooked up from imported Greek raisins or from sugar syrup mixed with patent dyes and potions. Perfectly genuine wine meanwhile was being imported from Spain, Italy and Algeria in vast quantities – which some Bordeaux merchants were not above labelling up as "claret" for export. As they read of the destruction of France's vineyards in their newspaper, wealthy Londoners and Bostonians began to ask exactly what it was they were drinking. A shadow of suspicion fell on French wines which would take decades to lift.

30

❧

The talk in the anguished 1870s had been of "defense" against the aphid. Since the Congress of Bordeaux, in the Midi at least, it was of "reconstitution." It would be a colossal undertaking. By 1884 1,000,000 hectares of France's vineyards had been destroyed, and 664,000 more were in the grip of the parasite and dying. At the same time only 53,000 hectares had been replanted with American vines, either direct producers or grafts, 30,000 hectares of which were in the department of the Hérault.

In Paris the balance of opinion was beginning to shift. Professor J.-B. Dumas, president of the Superior Commission, was terminally ill (the distinguished chemist died in 1885, to be replaced as head of the commission by Louis Pasteur). While a phalanx of sulphurizers still held sway on the august body, Professors Planchon and Balbiani energetically advanced their competing biological solutions. The commission's report for February 1884 showed the first crack in the insecticide-or-nothing line. It began with a review of the latest submissions from prize-hopefuls. Ten years on, nothing had changed.

"Inventors from all parts of the Old and New World have continued to propose procedures," Dr. Ménudier, the prize subcommittee's *rapporteur*, informed the minister of agriculture. Unfortunately their remedies were "completely lacking in any kind of authentic experimental proof." No surprise there, but perhaps M. le Ministre would like to know that the latest batch of proposals included application of the secretions of *escargots* dissolved in salt water, poisonous plants, toxic salts, certain acids, "electricity" and "the blessing of the seeds of grapes before putting them in the ground." The prize had become a running joke, kept going for the amusement of the phylloxera commission's *savants*.

"But much more happily, growers have not despaired," continued Dr. Ménudier, himself a proprietor of the Charente and perforce experimenter with alien plants, "and the results of the fight on many fronts affirm that our vines can be defended with a considerable chance of success by using sulphur of carbon and potassium carbonate." Now he came to it: "But is it not our duty to add to insecticides a weapon which, having been unpityingly rejected by many, is now being justifiably welcomed with open arms? We speak of American vines."

In the Senate and National Assembly, the argument rejected in 1879, that "reconstitution" as well as insecticide should receive state financial backing, had loudly resurfaced. The sulphurizers would not surrender quietly. According to M.R. Dezeimeris, counsellor general of the Gironde, vine-grower and American enthusiast, figures sent to the Ministry of Agriculture were being deliberately falsified. "Every time it was a question of 'defense' [insecticidal treatments], the head of the Gironde Agriculture Service judged that evaluations of 'success' coming from the municipalities were too low," he wrote, "and altered them upwards. Every time the figures related to 'reconstitution' they were marked down."

The immediate problem was getting hold of the new vines. The first commercial nurseries licensed by the Ministry of Agriculture had appeared in the phylloxerated Midi. Agricultural societies did what they could; that of the Gard for example began distributing young vines to small owners in 1879 in proportion to the amount of taxes they paid. There were not nearly enough to go round. In scarcity there was opportunity. Along with the prefecturial inspector, a new figure of peasant demonology arose – the itinerant vine-seller, touring the country markets with a wagonload of his wares, their actual provenance often wreathed in mystery. Why not buy now? They would only be dearer next week.*

Victor Pulliat wrote: "The enormous profits that can be made out of American vines has meant the sudden appearance of traffickers who buy and sell vines as they might umbrellas or socks. One or two years ago these self-styled 'gardeners' knew as little about American vines as they did about native ones."

Diligent *viticulteurs* could reproduce their own vines by propagation. In a rising market the young plant was of far more immediate value than its future product. Old soldiers were hired to act as vine-herds, and signs were put up warning of "wolf-traps," as much to drive the speculative fever as to deter rustlers.

Slowly, slowly, reconstitution gathered pace. When the Beaujolais was officially declared phylloxerated in 1880, the import of alien vines became legal. A handful of bravehearts resolved to try them. In 1881, Émile Duport at Saint-Léger on the east slope of the Côte de Brouilly put in Gamays grafted on American roots. When they gave their first harvest in 1884, Duport could proclaim: "The phylloxera – this time we are holding the line."

It was Pulliat who became the *américainiste* apostle of the Beaujolais. On Sundays at Chiroubles he held after-church grafting classes, treated at first as if they were the prayer-meetings of a heretical cult. Very few dared attend. Anyway, only the rich could afford it – for the small producers the cost of replanting with grafted vines was prohibitive, the three-year

---

* M. Lasserre, director of the nursery at Agen in Lot-et-Garonne, noted in 1886: "The demand for *les américains* has been transformed into a torrent. Everybody wants to plant them, all the former mockers beg for vines and complain angrily if there are none. Rooted cuttings which at the beginning of the season were being sold for fifty-five a hundred are being sold on for thirty-five francs."

gap in production unsupportable. As insecticide-dosed vines began to fail, the poorest just abandoned their plots and sought survival for themselves and their families as hired workers. The view from the heights of Beaune meanwhile was getting less lofty. As in the Médoc, the grand proprietors stayed firmly chemical. Those who unpretentiously turned out *vin ordinaire* were giving up the fight. Their vineyards on the hilltops and marginal lands were not worth defending. The argument raged through the first half of the 1880s. The insect meanwhile was doing the work of the American partisans for them. By 1885 half of the vineyards in the communes around Beaune were dead or infected beyond chemical redemption.

A revolution from below was coming. In 1886 the town's newly founded Société Vigneronne, its thousand-strong membership recruited from "the working class," began lobbying intensely for the edict on foreign vines to be lifted. A monster petition was raised. The official Vigilance Committee remained utterly opposed. The conspiracy theories resurfaced: these "pretended anti-phylloxera committees live off our misfortunes," said a correspondent of *Le Petit Bourguignon*, as apparently did the "*piquette*-merchants who accumulate fortunes with their dry raisin confections."

The society's secretary, M. Latour, toured the vineyards of Beaujolais, including Victor Pulliat's *champs d'expérience* at Chiroubles, to see for himself the replanted rows of Gamay and Pinot on American rootstock about to bring apparent salvation for all. On his return to Beaune, Latour shrewdly steered a middle course. His report recommended keeping as yet uninfected Burgundian vineyards going with insecticide, while uprooting the dead plantations and replanting with grafts.

Many American vines, hybrids and grafts had in fact already been acquired illegally from unscrupulous dealers in the Midi. On 15 June 1887, the Côte d'Or was officially declared phylloxerated and the ban was lifted.

To Victor Pulliat the vine-smugglers were heroes. "Just two

or three years ago, a few brave innovators planted American vines in the neighborhood of Beaune in the face of their neighbors' curses and threats," he wrote. "Now these newcomers enjoy the right of citizenship in the citadel of vintage wines, where all the members of the Société Vigneronne proclaim themselves to be ardent *américainistes.*"

In the fields of France peasants were voting with their grafting knives. In Paris too the argument was being won. At the end of 1886 a caucus of representatives from the south moved a draft bill in the National Assembly for an outright lifting of the *impôt foncière* (land tax) "in the departments ravaged by the terrible insect." The minister of finance declared it would bust the national budget, costing twenty million francs a year. Four months later the bill was re-presented, inviting a more affordable four-year (the time it would take a young vine to come into production) tax moratorium on land planted or replanted with new vines.

The *sulfuristes* on the Superior Commission had lost the will to argue. Professor Dumas was dead. Maxime Cornu had converted to the American cause. The official body had been merrily declaring phylloxerated departments open to the import of alien vines and doling out medals for grafting ever since. Its report of May 1886 especially extolled the *vignerons* of the Hérault who, "after twenty years of struggle, have remade a vineyard in this one department alone bigger than the whole of the United States, California included . . . The administration is making every effort to assist and develop the movement of reconstruction." The bill was passed, and was ratified by the Senate in November 1887 without alteration. It became law on 1 December.

The triumph of the Americanizers seemed total on both the scientific and the political front. Professor Millardet's experiments had continued. Having raised to maturity a hybrid rootstock that at last appeared to work in chalky soils, the Bordeaux botanist could confidently announce: "1887 will mark a date

never to be forgotten in the history of our wine disasters . . .
Due to the hybridization of our European vines with various
American vines, we are from this day forward absolutely certain of obtaining, from the first generation, either rootstocks
of an assured resistance and soil-affinity or direct producers,
resistant to the phylloxera and capable of producing completely
palatable wines."

Professor Millardet was wrong on both counts. France's
wine disasters were not over yet.

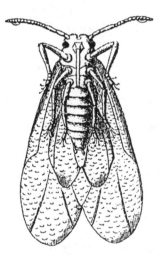

31

The soil of France, charged now in the *zones phylloxérées* with more than ten years' worth of reproduction by teeming American insects, did not welcome American vine-roots with universal greeting. Where they had been planted in calcareous (high calcium carbonate – limestone) soils, it was observed that spring leaves rapidly yellowed, while the vines remained green. The plant might struggle on, but would eventually die. The condition had been observed before in vinifera plantations on chalky soils, especially after heavy spring rains, but very rarely proved fatal. It was known as "chlorosis."* At first the phylloxera was blamed, but these were vines – planted as both rootstocks and direct producers – that had shown exemplary resistance to the underground devastator on other types of soil.

Limestone soil was predominant in key wine-growing regions of France. The brandy-producing Charente, for example, where the geologist Henri Coquand had shown in 1860 that the quality of the thin, acidic wines from which the best cognac was distilled was directly related to the limestone content of the soil in the Grande and Petite Champagne – the higher the better. The white wines of Chablis, Vouvray and Saumur originated on calcareous soil. There were tracts in the Midi and Burgundy and in the Gironde, in the chalky outcrops around

---

* Known variously as lime chlorosis or iron chlorosis – caused by too much calcium uptake by the roots, which leads to iron deficiency and thus insufficient chlorophyll formation and leaf yellowing.

Loupiac and Sainte-Croix-du-Mont. In the as yet unphylloxe-rated Champagne, the soil around Reims and Épernay was up to 60 percent chalk.

Just as in the first stages of the aphid's attack, the malady declared its presence with bafflingly complex symptoms. Certain American varieties remained strong and healthy before being decapitated to become a rootstock. After the grafting of a vinifera, however, the leaves of the transplant became chlorotic. But it was also observed, conversely, that the leaves of a transplanted vine were sometimes less afflicted than those of some American rootstocks themselves, as they were grown to maturity before grafting.

This was not altogether surprising. Just as American natives had evolved resistance to the animate parasite, Europeans had adapted to the inanimate "stress" of high lime-content soils. The first Vitis labrusca–derived varieties to go into chalky ground – Clinton, Taylor, Cunningham and Concord – proved woefully vulnerable. And when from 1878 onwards Vitis riparia was advanced as the wonder rootstock to beat the aphid, these varieties – solonis, Jacquez, York-Madeira – also proved all too susceptible to the hateful chlorosis. In the Charente there was uproar. Vineyards so expensively reconstructed on American stocks were in ruins. Chalky hillsides, beneficial for the quality of the grapes, were abandoned for the bottomlands where the lime content was not so high. The landscape was emptying. Writing in the 1930s, the agronomist Henri Hitier recalled a journey through the Charente fifty years earlier. "Whole villages were just abandoned," he wrote, "presenting a scene of desolation and ruin that we would not see again until the end of the Great War when we could go back into the villages of our liberated regions."

It would take time, but just as with the living aphid, there would be a biological solution to the inanimate hostility of the soil. On his viticultural tour of Missouri in 1873, Professor Planchon had examined in Dr. George Engelmann's study in St.

Louis a curious wild vine with leaf-galls containing mummified phylloxera. It had been collected three decades before in the Béxar district of Texas (then part of Mexico) by the explorer-botanist Jean-Louis Berlandier on his tempestuous mission for the Geneva Philosophical Society. But the dried sample from his "herbarium" never reached Switzerland.

By a circuitous route – Berlandier's Mexican widow had sold it to a U.S. Army officer, who sold it for $500 to Asa Gray in Washington – this botanical scrap, known simply as "Item No. 3115, Herbarium Berlandierianum Texano-Mexicanum," had arrived with some handwritten notes to be filed away by Dr. Engelmann in St. Louis as a curiosity. Berlandier himself was long dead. His field notes described the wild vine he had found growing profusely on the arid, chalky uplands of Texas, where it was known as the "mountain grape."*

Seeds of the mountain grape had been sent to France in 1874 from the Gulf of Mexico by the eccentric Texan nurseryman Gilbert Onderdonk, to the care of Gustave Foëx at the Agricultural School of Montpellier. More arrived the following year to be planted by the experimental *viticulteur* L. Giraud of Nîmes. Word began to spread. Rooted cuttings sent from a nursery at Indianola, Texas, were planted by M. Davin, a doctor of Pignans in the Var. The good doctor observed his exotic vines growing "beautifully" on a "white, calcareous outcrop where no other vineyard could survive." There were problems – propagating by cuttings proved persistently difficult, and those that worked were very reluctant to take root. Nevertheless, science was waking up to the potential of the mountain grape.

Planchon compared the desiccated samples and the drawings he had retrieved from Berlandier's herbarium with the "monticolas" now growing at Montpellier. They were the

---

* It was also called little mountain grape, sweet mountain grape, sugar grape, winter grape, fall grape.

same. In a taxonomical coup, in August 1880 he renamed the species "Vitis berlandieri Planchon," announcing the move in the proceedings of the Academy of Sciences.

The hybridizers saw the possibilities of the strange plant that liked hillsides and chalk. Alexis Millardet took up camp at Giraud's vineyard, methodically testing the phylloxera resistance of the "new" lime-loving vines according to his clever scale. It was excellent. The problem lay in propagating the magic mother, Texas-raised berlandieri, wholesale to make French-born clones. They just would not take.

Millardet's solution was to hybridize the American plant – with a Chasselas at first (a cross dubbed 41B), then with other vinifera varieties, then with other American species. Gustave Foëx at Montpellier impregnated Cabernet Sauvignon with berlandieri pollen – the resulting hybrid was dubbed 333EM (École Montpellier) – and made more Americo-American and Franco-American crosses. The *hybrideurs* were looking for the holy grail, a single "direct producer" that would resist phylloxera, be immune to chlorosis, be easy to propagate and bear fruit without ignoble taint. It would be a long and frustrating search.

The grape-growers of the Charente and the big distillers who were their customers could not wait that long. Many farmers had their own stills, and made a first distillation to be sold to the grand houses of Cognac. Big and small operators were buffered by the fact that brandy production requires the carrying of large stocks for ageing and blending. But those reserves would run out. In desperation the Martell company tried distilling 150 casks from the fermented juice of American vines already planted in the Grande Champagne.

The product was anything but grand. In 1886 the Central Vigilance and Study Committee of the Charente-Inférieure, supported by the Hérault's battle-worn Central Agriculture Committee, petitioned the officials in Paris to send "one or several delegates to America to look for varieties of native vines which would grow strongly and without chlorosis on chalky lands identical to ours and at a similar latitude."

Eugène Tisserand, permanent secretary at the Ministry of Agriculture, agreed to the proposal, deputing a twenty-eight-year-old professor at the National School of Agriculture of Montpellier for the mission. His name was Pierre Viala, and he had been born at Lavérine in the Hérault. Fourteen years after Planchon had made his congenially civilized tour of the Mississippi-Missouri valley, the young plant pathologist's *mission viticole* would take him to some much wilder places.

It began as Planchon's had done, with a trip to Washington D.C., but this time the French visitor was more concerned with rocks than plants. Viala consulted the U.S. Geological Survey, which gave him "hope of finding vines growing in pure limestone." He pored over maps looking for such a stony promised land. His coast-to-coast journey took from June to December 1887, with all sorts of suitably wild west adventures along the way.

Planchon's key encounter in America had been with the insect expert Charles Valentine Riley. Pierre Viala's was with a gardener-botanist, Thomas Volney Munson, who throughout his life conducted an extraordinary love affair with the fruits of the vine. Born in Illinois in 1843, the same year as Riley, in 1870 he had been appointed professor of natural science at the University of Kentucky in Lexington, close to where John-James Dufour had founded his doomed "First Vineyard" almost sixty years before. Perhaps it was the Swiss pioneer's ghost that made Munson declare: "The grape is the most beautiful, most wholesome and nutritious, most certain and profitable fruit that can be grown." He tried growing his beloved grapes, including native labrusca and European vinifera (which died mysteriously) in Lincoln, Nebraska, but even his native varieties were defeated by the cold winters and the grasshoppers. He found a more benign spot at Denison, Texas, sixty-five miles north of Dallas,* founding a nursery and beginning an ampelographical collection.

For years Munson had traversed Texas, Mexico and beyond, driven by his curious vine-hunter's passion, and in 1885 he exhibited in New Orleans a splendid herbarium of all known American species. Viala sought him out. There were plenty of berlandieri in Munson's nursery, which he had managed to propagate successfully from cuttings. Where might Viala find the mountain grape growing wild? Would the soil really match that of the Charente? Thomas Munson knew exactly where his French visitor should look. "I had had to spend a long time searching for the different wild vines in forests," Viala wrote. "Only in the north of Texas did I think I might discover the kind of terrain that I was looking for. When I consulted Mr. T.V. Munson, he showed me where I might find in central Texas chalky soil where native vines grew green and strong."

There indeed he found the miracle vine, growing cheerfully in the dry, dazzlingly bleached hills around Belton.

Pierre Viala's report *Une Mission viticole en Amérique* was published in Paris in 1889 as a big, glossy hardback. In a still very crowded market for vine-malady tomes, it was a big seller. The young botanist became famous. He could not recommend the lime-loving, phylloxera-resistant berlandieri highly enough, although he had to admit certain difficulties in propagation and getting samples to take root; but he "hoped that it would be possible to discover more amenable varieties."

The success of Viala's book made Thomas Munson's export

---

* As J.-L. Berlandier had predicted fifty years earlier, the Texans had indeed made a viable little industry out of native vines. By 1860 German immigrants around Fredericksburg had tamed the wild Mustang grape. In 1870 the Sidney Borden Winery in Sharpsburg was shipping a white wine called "Sharpsburg's Best" and a red called "Rachel's Choice," named after Borden's wife. Italian immigrants in the Dallas area took up the challenge, making "passable" reds.

business boom, just as over a decade before Planchon's publication of *Les Vignes américaines* had filled the railroad boxcars leaving Missouri for the Atlantic ports. Berlandieri was the new wonder-stock. From 1888–89 large quantities of bundled vines began to be shipped from Denison in sealed crates via La Rochelle to the officially phylloxerated Charente, where the import ban did not apply. The press exulted. Chlorosis and the aphid had been beaten together.

The rejoicing was premature. Berlandieri, as the first experimenters in the Hérault had found, proved intractably difficult to propagate. Those who managed to do it found the resulting plants excellent in terms of soil adaptability – and they would take vinifera grafts when it was done with care. But there were just too many failures. In spite of government incentives, French nurserymen and cottage-industry propagators simply gave up trying. Pierre Viala's heroic star fell to chalky earth.

Prosper Gervais, a disgruntled vine-grower of Montpellier, later summed up French feelings on the Texan vine's rise and fall: "Although many spoke of it, no one had seen it. Berlandieri was like some terrible Oriental divinity, remaining hidden from believers behind the heavy hangings of the sanctuary. But Viala, berlandieri's high priest, could not rest until he had torn away the veils and made the God appear. Well now we have seen Him. Now we know everything or nearly everything about Him. And I have to say that very few of us have been converted . . ."

The wreck was not total. The hybridizers had already been tinkering with the mountain grape for over a decade. Professor Millardet's Chasselas x berlandieri cross – 41B – was unveiled at the Mâcon Wine Congress of 1887 amid general excitement. Georges Couderc set up shop in the Charente, with an experimental vineyard called Tout-Blanc near Marville where he tested huge numbers of hybrids as potential rootstocks. They bloomed green amid the arid white landscape. Millardet himself arrived from Bordeaux to take up the challenge of the lime-rich, phylloxerated soil, successfully grafting Charentais vines such as Folle-Blanche and Saint-Émilion onto his sturdy 41B hybrid rootstock.

The businessmen of Cognac meanwhile launched their own front in the berlandieri war. In 1888 the Comité de Viticulture commissioned the twenty-three-year-old Louis Ravaz, a star pupil of Foëx and Viala at the National School of Montpellier, to direct the reconstitution of their devastated vineyards. As Millardet had done with phylloxera resistance, Ravaz worked out a table of tolerance to soils of varying chalkiness and per-formed further work on the affinity of venerable varieties with the flashy new transatlantic rootstocks.

The generalship in the battle against the aphid had long since passed to the hybridizers and rootstock technicians. Jules-Émile Planchon's voice was by now muted although he still lectured and propagandized especially against the winter-egg theorists. In spite of the urgings of his family he refused to press the government for the "phylloxera remedy" prize money which many thought should be his. On 1 April 1888 he died suddenly at his home outside Montpellier, aged sixty-five. A subscription was announced to provide a pension for his widow and to raise a suitable memorial.

By the last decade of the nineteenth century the biological tools to put the vineyards together again – whatever the *terroir*, whatever the revered variety that had to be kept in production – existed as prototypes. It was now a question of mass produc-tion of grafted vines – and, as important, a continuing process of persuasion that they were indeed the answer. Pierre Viala propagandized intensely, setting up *champs d'expérience* across the country where even the most intractable doubters could see for themselves.

The proof that it worked was in the taste. A *grand cru* on American roots was still a *grand cru*. If it was not, no one was saying so quite yet.

32

"Reconstitution" rolled out across the vineyards of France following the pattern of the aphid's devastation. Where the invasion had first begun in the Midi and the south-west, so in those regions a quarter-century later was the transformation of the vineyard most advanced. In the Hérault, for example, 166,500 hectares had been transplanted with American vines by the mid-1890s. In the Gironde, out of 138,000 hectares, 40,000 had been transplanted and 60,000 were reported as infected. In France as a whole, according to Ministry of Agriculture statistics, 663,000 hectares had been transplanted of a total surface of 1,748,000 hectares by 1895. Ungrafted American vines planted in the chaotic first years for want of anything better were beginning to be abandoned, supplanted in places by the new breed of "hybridized" direct producer.

The whole process known as *l'encépagement*, the choice of which varieties went where, was altering. The old varieties of Mourvèdre in Provence and Négrette in the Tarn were abandoned when they proved difficult to graft. The once Pinot-dominated vineyards of Sancerre were remade on grafts of Sauvignon to make a white wine that would be fêted in Paris. La Folle Blanche, the traditional grape for making brandy, was supplanted in the Charente by Ugni Blanc. Huge quantities of Aramon were planted in the south.

As the little plots on the hills were abandoned, the new vineyards of the plain were remade on lower densities of planting. Vines were now raised on wires in marching avenues for ox-

and horse-drawn wagons, ploughs and machinery to pass between. Grafted vines required greatly increased use of fertilizer. The mildew and yet another imported American fungus known as black rot which thrived in cool, wet summers demanded constant chemical treatment. In spite of animal-powered modernity the new vineyards were more capital- and labor-intensive than those the aphid had obliterated. The traditionalists hated them. The south had become a "wine factory." Its production, however, was abundant.

The phylloxera had not gone away. Chemical defense still held the line where the value of the wines (and the disdainful elitism of proprietors) allowed it. In the Haut-Médoc for example some proprietors kept up the work of injecting the soil until 1914 and beyond – until laborers went off to the front and the chemical factories were set to the manufacture of shell propellant and explosive.*

Other proprietors began working from the early 1890s onwards along the lines prescribed by Professor Millardet, hero of the battle against the mildew, energetically planting vinifera grafted on the professor's hybrid riparia x rupestris, a rootstock specially optimized for the Médoc's high-silica soil. Others like Latour began a slow process of changeover, replanting with grafts while keeping old vines going with insecticide where they could.†

Reconstitution in Burgundy ran to a similar pattern – insecticide for the grand, rip out dead vines and start again for those who had nothing left to lose. In spite of the populist triumph of

* In the First World War the Germans unsportingly used glycerine derived from grape juice to make high explosive.

† The 1911 *Encyclopaedia Britannica* notes: "These methods were chiefly advocated in vineyards of the first class, where it was worthwhile to spend a good deal of money and labour to preserve the old and famous vines . . . Some good judges attribute the peculiar and not unpleasing flavour of certain clarets of 1888 to means thus adopted to kill the phylloxera."

Beaune's *vignerons* in 1887, there were huge uncertainties. Those who had already illegally planted American direct producers found the fermented product of their grapes a poor ambassador. Grafting remained a mystery. Three skilled grafters were brought north from Villefranche and set up courses in the strange new art. Experimental *champs d'adaptation* were established at Beaune and Dijon and out in the countryside, where proprietors offered the use of their land in return for getting their hands on the wonder-vines. The pace was slow. Only a hundred hectares were replanted by 1890. The following year the figure had gone up tenfold – but still growers proved reluctant to abandon even dying vines, "not so much because of doubts of ultimate success but because of the high price of the plants," reported the departmental vigilance committee.

In September 1891 the now-famous Pierre Viala addressed Beaune's autumn Wine Congress. He would talk "vine-grower to vine-grower," he said (his father was a *vigneron*), and leave science to the experts. In fact he gave his audience a highly philosophical discourse on the lessons of the struggle with the aphid, arguing that the dogmatic battle between sulphurizers and Americanizers was over an illusion. The rifts and squabbles had slowed down the work of salvation. What mattered was the right solution in the right circumstances. Insecticides would be effective in all the varied soils and terrains of Burgundy, Viala said, especially if treatment were started at the beginning of the invasion or when the vine was still vigorous. A vine could theoretically live on "indefinitely" with its regular dose of chemical. But it would become uneconomic for all but the highest-value wines. On the *vin ordinaire*–producing hilltops, redemption by either chemistry or reconstitution was problematic. They would have to be abandoned.

The triumphs and disasters in other regions of France had shown the way. As for rootstocks, there were only a small number to consider: Vitis riparia, solonis, Jacquez, rupestris, and

Vialla.* In the one tenth of the soil of Burgundy that was calcareous, the Jacquez and the solonis would die of chlorosis. As he spoke, Viala was awaiting the final results of Louis Ravaz's experiments in the Charente, but he and his hybridizing colleagues were confident that at least some of the chalky vineyards of the Côte d'Or would bloom green again.

Something bigger was at issue – that old intangible: what did "grafted" wine taste like? The vines of those Beaune vignerons who had embraced grafting four years earlier were just maturing to bear grapes, but their wine was still a long way from being ready for drinking. For a good Burgundian to drink anything from the Médoc or the Midi, grafted or not, would be an act of treachery. It was completely untrue, Viala told the conference, that circulating "sap" imparted fugitive tastes and odors to the grape, as the mischievous and uninformed were suggesting. The biology of the vine did not work that way. Those who had doubts should look at the evidence from the Beaujolais, whose growers could testify that the vines they had grafted eight to twelve years before were giving wines just as good as or better than ungrafted vines of the same age. Experiments in the Blayais, in certain forward-looking *grands crus* in the Médoc, and testimony from the Libourne and Saint-Émilion proved it, Viala insisted. The quality of their wine, the untainted purity of its taste, had been secured on American roots.†

Not everyone agreed.

* * *

* Vialla: a variety of labrusca, which somehow survived the phylloxera's attacks, reputedly raised from a seed of Clinton obtained by Léo Laliman from the Bordeaux Jardin des Plantes. He named it in honor of Louis Vialla.
† In the spring of 1894 the owner of "La Goutte d'Or" vineyard at Meursault was awarded a gold medal at a Paris wine *concours* for white Burgundy from a three-year-old grafted vine. The minister of agriculture later arrived in person to present the Agricultural Order of Merit. According to Robert Laurent: "It was a decisive moment, serving to sweep away the last doubts of the proprietors of fine vines."

There was one great wine citadel of France yet to fall. In summer 1890 the aphid at last reached the low, chalky hills of the Champagne. There was the usual uproar. Church bells were rung in alarm. "Every newspaper has announced that the phylloxera has been discovered near Épernay," said the *Journal d'agriculture pratique* on 14 August; "the exact news is confused – the truth seems to be that the phylloxerated spots are in the commune of Tréloup in the Aisne   but close enough for the vineyards of the Champagne to be endangered. Extinction treatment will be applied immediately to the infected vines."

The region's vine-growers had had years to prepare, but the story was familiar: denial and resistance to the eradicators. The stern M. Gouanon, inspector general of the Service Phylloxérique, moved in with pickaxes and petroleum. The Tréloup spot was burned, as "proprietors, at first recalcitrant, began to understand that their interests lay with the greater good and allowed access to the investigators," according to the *Agriculture pratique*'s correspondent. If energetic action was taken, extinction of the plague was possible. The Champagne on its northerly raft of chalk had a vineless cordon around it. The great houses of Chandon, Gallice, Werlé and Pommery between them very quickly raised a fighting fund of twenty thousand francs, subvented to the prefecturial professor of agriculture.

A week later the aphid was found in the department of the Marne itself – at the little village of Vincelles, forty kilometers due east of Épernay. M. le Comte Gaston Chandon knew what to do: he pressed a wad of francs into the hands of M. Piot-Husson, the startled proprietor, and promptly set his vines to the torch. The gesture made Chandon a local hero. "It is still a matter of the greatest regret," editorialized the *Journal d'agriculture pratique*, "that the government did not act thus in 1871 destroying centers of infection in the Midi." The gesture of the Épernay *négociant* was to be greatly applauded – but the writer pointed out the dangers to the *champenois* if the hoped-for extinction of the aphid by burning should fail. The patchwork vineyards of Pinot Noir and Chardonnay, owned by a myriad of small farmers who sold their produce to the grand houses, were planted *en foule* – closely packed – and cultivated manually with traditional tools and age-old methods. Vines were crowded in thirty thousand to the hectare, five times the typical density in the Bas-Languedoc. Dosing them with insecticide would require an ocean of carbon bisulphide. It was burn or perish. An anti-phylloxera syndicate was formed, recognized by prefecturial edict, in July 1891.

The attempts of official eradicators to move in on suspect plots met the same peasant intransigence as everywhere else in France. The common front of *vignerons* and their grand customers in Reims and Épernay quickly fell apart. According to an investigation by the Paris magazine *L'Illustration*: "The administration aspired to expropriate contaminated land before the syndicate committee had been elected. When, after a passionate campaign, the elections were held on 12 August 1891 there were two lists – one composed of *vignerons*, the other made up of members of the Syndicate of Commerce of the Wines of Champagne." There followed six months of squabbling and gerrymandering. The prefect of the Marne pushed through by "subterfuge" the constitution of a committee weighted in the big houses' favor.

"From that moment the syndicate completely lost the confidence of the small proprietors," according to the magazine's report; "the twenty-five elected *vignerons* resigned, and if in some places uprooting and burning was indeed carried out – at others there was absolute opposition. Some firebrands were preaching: 'Would you tell the *vignerons* they must let themselves be pushed around by this illegal syndicate which, under the pretence of destroying the phylloxera, wants nothing but to snap up their land for the profits of the big wine merchants?'" It was the same mix of intransigence, ignorance and conspiracy theory that had brought out the gendarmerie in the Côte d'Or a decade before.

A certain M. Demont of Reims meanwhile dusted off an old theory. Champenois vines had become "degenerate" because of centuries of "abnormal reproduction" by vegetative propagation, he insisted. American vines resisted the aphid because they themselves originated from sexual crossings – although now they too were beginning to fail. His solution was to raise a revitalized race of new French vines from seeds. "Will this fountain of youth succeed?" asked M. Crépeaux, *L'Illustration*'s reporter. "We fear not." Crépeaux was right. What would eventually save Champagne was vinifera vines grafted onto rootstock that would resist both the insect and the fatal chlorosis caused by chalky soil.

As in the Charente, it was the big houses that had the money and the staying power. With the failure of eradication, however, some resorted to insecticide. Gaston Chandon's brother Raoul, president of the Association Antiphylloxérique d'Épernay, applied massive doses of carbon bisulphide at the company's own vineyards near the town. The insect's comparatively slow advance gave him the time to note the influence of the chemical on vines of different ages and in different soils. It worked well in light, permeable soils, but older vines were killed. As in Burgundy, it was becoming clear that chemical defense was an expensive dead end. Two questions had to be answered. Could the traditional growing methods of the Champagne work with grafting, and would grafted vines somehow taint the end results?

Raoul Chandon went to the Midi to study the work of reconstitution carried out under the guidance of the Montpellier School of Agriculture. On his return to Épernay he eventually established more than a hundred experimental plots. M. Manzade, a Montpellier graduate, was recruited as chief researcher with the mission to find the best type of grafting for the Champagne.

Moët's École de Viticulture, established in 1899 at Fort Chabrol, Épernay, would train a new army of *vignerons* in the art of grafting. The most widely used rootstocks would be berlandieri hybrids and riparia-rupestris crosses, although other less suitable varieties – hybrids of solonis, of rupestris and of riparia – were used by humbler growers.

The vineyards of the Champagne were not completely "reconstituted" until the 1920s. In other parts of France, the Romanée Conti in Burgundy for example, commercial ungrafted vines lingered on for much longer. Across the country diehards and eccentrics kept little plots of old vines alive with chemicals to serve a market of true believers – those who would not accept, whatever Pierre Viala had to say, that grafted wine was the same as that from vines the phylloxera had destroyed.* At the outset of the new century the *derrière-garde* found a controversial champion.

Lucien Daniel did not come from a wine-producing region, as his many enemies were keen to point out. However, the professor of botany at the University of Rennes did have some advanced views on "reconstituted" vineyards and their products. He first expressed them openly at the Lyon Agricultural Congress of 1901. The emperor had no clothes, said the Breton botanist – or rather, wine from grafted vines was not as good as wine from those that had gone before. Grafting might indeed increase a plant's grape production, as its champions proclaimed, yet, so they said, the fox-tainted rootstock had no influence on the quality of the scion's fruit. How could this be? The grafters could not have it both ways. The rootstock could and did modify all the characters of the transplant, he insisted; not only certain morphological characters but also resistance to diseases, chemical composition of fruit-pulp and the wine itself. It was quantifiably worse. His reception was hostile.

Nevertheless the Ministry of Agriculture commissioned Pro-

---

* For example, in 1907 Château Latour signed over futures in five annual harvests to a Bordeaux *négociant* with a clause in the contract insisting that American "transplants" be banished from the vineyard and only French varieties be employed.

fessor Daniel to inquire further. In 1902 he concluded that "grafting combined with a more intensive culture, is largely responsible for the disasters which have overtaken the wine-grower: too much wine and wine with unpleasant underlying tastes." Daniel's doom-laden pronouncements may have suited the hybridizers with their cheap and cheerful solution. He was shunned, however, by those who had expensively followed the grafted path to redemption. There was a conspiracy of silence, he alleged. His countrymen had lost their sense of taste. Why, he knew a fellow professor who drank American wine and never complained of the "taste of fox." When nobody else would listen, he eventually found refuge in the columns of *The Times* (an exhibition of French wines in London provided the excuse), blasting those who had bewitched France with the "dogmas of reconstitution based on errors of physiological theory and blunders in cultural practices." "Wines from grafted plants have by no means the constitution of the old wines," he informed alarmed *Times* readers:

> They are quicker in becoming fit to drink, but quicker also to go off. The wines of certain crus have lost their distinctive characteristics. They are no longer comparable to the old wines. That is the ugly fact ... Thus in the device of grafting and the vicious practices which it entails you have the explanation of the evil pass to which the vine-growing industry has come ... At all costs buyer and seller must get the good wine that they used to get. There is only one way, have done with grafting; to restore them by returning to the deserted hillsides; to cultivate the vine once more in accordance with the methods and experience of centuries ...

Professor Daniel was either very foolish or very brave. To say over a few agreeable bottles that things were not as good as they used to be might be considered reasonable. To say so about French wine in a foreign newspaper was an act of treason.

The professor was stripped of his Légion d'Honneur for his outburst.*

It was a very sensitive time. A new crisis was rumbling through the vineyards. At the grander end it was about exports – was claret still claret?, it was being asked in London gentlemen's clubs. Professor Daniel said it might not be. At the humbler end it was about too much wine, now pumping from the "reconstituted" vineyards of the south in a torrent that even the thirst of France would come to struggle to consume. In London it meant a new vogue for Scotch whiskey and soda. In southern France it triggered a popular uprising.

La cuisine viticole, as the French called it (the "cooking up" of phoney wines), had not gone away. In 1896 a Girondin winegrower named Albert Bonnet published a pamphlet pointedly called La Chaudière ("the boiler"), attacking unscrupulous merchants who switched foreign wine for French, who had invented a fairyland of phantom châteaux and who manufactured " 'médoc' out of water, a few drops of brandy and some sugar candy." The wine-switch was easy. Barrels on the Bordeaux quayside marked "Product of Spain" would turn up in Antwerp with the markings erased en route.

Impersonating a grand cru was more complicated. It meant phoney labels, phoney documentation and phoney corks. But profits could be great, and penalties were hardly severe. In June 1904 a Bordeaux court sentenced a fraudulent merchant to eight days in prison for passing something off as Château Latour '94. When the vineyard itself was nonexistent ("Château des Géants," "Château Médauc," "Château Latour-Massac"), who, other than the duped faraway consumer, was going to

* Soon after its London publication Le Progrès agricole et viticole ran Professor Daniel's article in translation, noting: "It is superfluous to comment further on this odious diatribe against the eminent viticultors who have worked tirelessly for thirty years to reconstitute one of the principal treasures of France."

complain? Frauds furthermore might be just as well conducted in the country where the stuff was drunk. The German port of Hamburg was notorious as a production line of exotic beverages. There was a raucous passing-off scandal in Belgium in 1903, and huge embarrassment at the World's Fair at St. Louis, Missouri, a year later when the American press made the exhibition of Bordeaux wines an excuse to expose any number of "fraudu lent" vintages. Perhaps it was revenge for the humiliations heaped on the beastly Catawba by the French jury at the Exposition Universelle almost forty years before. In Russia too, in spite of a love of all things French following the military alliance of 1894, there was a loud newspaper campaign against the alleged falsifiers of the Gironde. The Russian aristocracy could not be seen to drink German wine. They turned patriotically to vodka.

In London things were as bad, if not worse. It was middle-class newspapers that now delighted in stirring the fraudulent claret pot. When in 1903 prices for recent Bordeaux vintages seemed "too good to be true," the Paris correspondent of the *Daily Telegraph* could state as a fact that half of them were *vins ordinaires* passed off as *grands vins* – according to a mysterious informant in the Gironde. The *Telegraph*'s man in Paris could also reveal, a little late with the news, that the French revolution had caused "socialistic land distribution," and that the grand-sounding wines of Bordeaux actually came from a myriad of tiny proprietors who sold their product to the nearest château. What the peasant sold for £7 a barrel (220 liters), Château Margaux sold on for twenty – just for the privilege of applying its not entirely patrician label. Was this not the true scandal? There were many hundreds of wine-making communes apparently, of which "only ten were known." He had also been reliably informed that a great deal of champagne exported from Reims was actually cider from the Vosges. *Telegraph* readers were outraged. The writer of the "Paris Day by Day" column would qualify his shock revelations in subsequent articles, but the seed was sown. Generation of aspirant Englishmen would follow generation, obsessively fussing over a bottle's provenance, fearful lest they were being gulled.

The noted oenophile William Pheysey, wine-buyer of the Army and Navy Stores, waded into the controversy at the 1905 wine trade congress held in Liège. His "sense of duty" led him to try to "put an end to these abuses," he proclaimed. "The crime of false pretence is too prevalent and is now looked upon apparently as the custom of the trade . . . The name of a particular vineyard is treated as a generic term which can be applied by anyone as he thinks fit! Saint-Julien no longer comes entirely from the commune of that name – but is a concoction of wines from other places . . ." By now that was hardly much of a surprise.

It was up to the proprietors themselves to set things right, according to Pheysey. "Well-known growths" must be bottled at their châteaux, he suggested. The name of proprietor and shipper must be emblazoned on the label as a guarantee of provenance. The "syndicate of Reims" had striven to protect its territorial identity – why could not other wine-growing regions of France do the same?

"Wine" meanwhile was indeed being confected on the banks of the Thames – from sugar, raisins and other strange ingredients. Two enterprising Cypriots had founded the Vintners' Products company to make "British wine" from imported grape-must. It was all perfectly respectable. "Millions of gallons of wine were being made annually in London and he should not like to say how much of it was innocent of the grape," so the *Daily Express* reported the comments of a "West End wine merchant" in March 1905. These so-called "basis wines" were legal and tax-free. And why not? At least the drinker knew what he was getting, which he did not with so-called "foreign wines," most of which were confected "in Germany where spirit is cheap, sugar is cheap and water even cheaper," according to a pro-British wine trade circular of the period.

Consumption of Bordeaux wines plummeted. The high-Victorian manner of drinking only champagne during dinner returned. More adventurous strands of polite society and the officers' messes of fashionable regiments converted entirely to whiskey and soda. Some nostalgists blamed the new habit of the after dinner cigarette. Maybe so but claret's Edwardian eclipse was the aphid's legacy.

33

🍂

After more than twenty years of anguish and effort the vineyards of the Midi had been put together again. The costs had
been great, debts were pressing, but by the mid-1890s the reconstituted vineyards were producing a flood of wine for which
there seemed to be no end of thirst. In the industrial cities of
the north, in the Breton ports, they were guzzling the stuff –
men, women and children. The new-made latifundia of grafted
Aramon and hybrid direct-producers marching across the hot
plains were delivering a colossal 150 hectoliters of wine a hectare. If their alcohol content was weak (six or seven degrees)
they might be blended with the *demi-muids* (half-hogsheads)
of fierce Algerian liquors now coming in bulk by ship to the
port of Sète. Sugaring was a cheaper option.

In 1893, for the first time in fifteen years, the French national
harvest came in at more than fifty million hectoliters, 40 percent of it from the *gros départements* of Languedoc and Roussillon. The overall figure dipped thereafter due to mildew and
black rot, then burst back into super-abundant health. It was
too much – ten million hectoliters too much by 1901. In the old
days the surplus might be distilled as *eaux-de-vie* or industrial
alcohol; now that came direct from sugar-beet.

The price of wine collapsed. In the 1880s, a hectoliter of
Midi wine might have sold for thirty francs (forty francs on
average for the rest of France). It was eleven francs in 1900,
eight francs a year later, and still falling. A learned Russian
economist, Dr. Christophe Sergueev of St. Petersburg, came

to France to examine the causes of the unfolding crisis. He pronounced in the columns of the *Journal d'agriculture pratique* in 1902 that it was not just a matter of overproduction. The revenue lost when the phylloxera reigned, the costs of reconstitution, the costs of the new intensive viticulture, had laden *vignerons* with unsupportable debts. The need to repay them meant selling as much wine and as soon as possible. The glut had slashed the value of the vine-supporting soil itself. Growers could neither repay their existing debts nor borrow any more to survive.

The economic effects of the flood of cheap wine sloshed through every vineyard in France. The revenues of all "common" wine-makers plunged. Just as the vine-growers of the north had stood by and watched the travails of the south when it was in the grip of the aphid, so the bloated production of the "new" vineyards pointed a harsh lesson. What they produced was worth half of what it cost to make.

When the still ever-advancing phylloxera reached the Paris region around 1900,* those growers who continued to tend their historic vineyards around the capital simply gave up. What the *oïdium* had begun, the law of supply and demand completed. Ancient vineyards on the periphery – in Picardy, Normandy and Brittany – were abandoned. According to the historian Marcel Lachiver: "It was not the phylloxera which ruined the vineyard in many departments, it was the certainty that there was no means to fight against the stream of wine which railway tank-wagons were pouring into Bercy."

The south was in turmoil. Where once the phylloxera had been the enemy of the people, this time the hate figures were human. It was the fabricators, the fraudsters, the importers,

---

* By which time almost the whole of France was phylloxerated. According to the Superior Commission report of 1900 it had reached the Ardennes. In the center La Creuse "presented a white spot in the map colored by the phylloxera," but there were only five hectares of vines under cultivation in the entire mountainous department.

the sugarers, the big joint-stock companies that had set up in the sand-dunes of the south which could hoard or release their product and thus set the market rate as they chose. Prices for "common" wine rose fitfully with the poor harvests of 1902 and 1903, but collapsed again with the three ample *récoltes* that followed. The financial plight of vine-growers rapidly became worse than when they had been in the grip of the aphid. There was a puddle of money and an ocean of wine. Even giving vineyard laborers two liters each a day to drink could not drain it.

Who were to be the more reviled – the *pisteurs* who now snooped round villages on behalf of the wine-brokers looking for growers on the brink of bankruptcy, or the politicians in Paris who refused to come to their aid? A champion arose, a middle-aged café-owner, vine-grower and sometime actor from the small town Argeliers in the Aude. His name was Marcelin Albert.

A cabalistic formula gripped the imagination of Albert and his ever-growing army of followers. The total amount of wine on the French market was eighty million hectoliters a year. Five million of these came from Algeria, five million from Italy and Spain, eight million was "fabricated" from sugar and raisins, and two million was in the form of *piquette* – a total of twenty million hectoliters. French harvests from reconstituted vine-yards now ran at above sixty million (of which two million was exported – all of it expensive prestige bottles). France could not swallow all that.* Shut down the sugarers, said Albert, punish the waterers, ban the Spanish and Italian imports (but not Algerian – they were made after all by brave Frenchmen that the phylloxera had evicted), and production would match consumption.

---

* 56.7 million hectoliters per year were being drunk by 1905, 61.5 million by 1909, according to government statistics that assumed tax revenues on commer-cially sold wine represented three-quarters of the actual total. The rest was un-taxed home consumption by *vignerons* and their families.

The state had in fact done much already. Foreign wine had been subject to an import tax since 1892. Legislation in 1897 had at last taxed raisin wines at the same rate as any other and set stricter limits on "the fabrication, circulation and sale of artificial wines." The ancient *octrois*, the niggling, long-reviled taxes levied on wine consumption by local municipalities, were at last abolished by the law of 29 December 1900. The law of 1 August 1905 further defined and set penalties for fabrication. This was not enough for Albert. His campaign against the sugarers burst into life in 1905 with a refusal to pay taxes in his home town of Argeliers, north of Narbonne. "Vive le vin naturel!" the café-owner's supporters cried. "Down with the poisoners!"

A parliamentary committee chaired by the Girondin proprietor Georges Cazeaux-Cazalet had been charged meanwhile to investigate the acknowledged crisis. On 11 March 1907 they descended on Narbonne to listen to the representatives from disaffected communes. Albert turned up with a gaggle of animated supporters to unfurl a banner declaring "Death to the Fraudsters." The commissioners evidently shared his outrage. Their initial report concluded: "The wine crisis is not due to the overproduction . . . The production of natural wine is not enough to meet national consumption."

The revolt gathered pace. That spring a vine-growing commune in the faraway Pyrénées-Orientales declared a tax strike. Albert brought out a campaigning newspaper, *Le Tocsin*, and a marching song was composed – *La Vigneronne*. Schoolmasters, priests, shopkeepers and bourgeois proprietors joined the swelling band of protesters. Their slogans were a bizarre mix of demands for consumer protection and revolution – "Sugar to the café, water to the canal"; "Wine from grapes, bread from wheat"; "War, war 1789–1907"; "Satisfaction or revolution."

The numbers of demonstrators meeting on Sundays in the towns of south-central France grew rapidly through the spring and early summer of 1907. From a gathering of six hundred at Bize on 31 March, eighty thousand besieged Narbonne a month later where Albert dramatically declared the formation of a "committee of public safety for the defence of viticulture." From that moment the movement became a mass populist uprising against the *fraudeurs*, the "beetroot-syndicate" of the north and the government in Paris, supposed to be their stooges. At Béziers on 12 May, a crowd of 120,000 people gathered to hear Dr. Ernest Ferroul, the tempestuous socialist Mayor of Narbonne, proclaim an ultimatum to Paris. If the government did not meet their demands by 10 June, a universal tax strike and mass municipal resignations would follow.

On 16 May the police barracks at Béziers was attacked and its files triumphantly dumped in the streets. The *démarche* of the countryside on Montpellier two weeks later was enormous. Special trains were run at reduced rates, and Mgr. de Cabrières, the bishop of Montpellier, ordered the churches to stay open as overnight refuges for women and children. On the evening of 9 June a crowd of over half a million listened to the prophet of *vin franc et natural* orate theatrically as the hours until the ultimatum ticked away – pleading the while that their great moral crusade should not turn violent. "The fraudsters to the guillotine!" roared the crowd. "To Noumea [the prison colony in New Caledonia] with the fabricators!" Off-duty soldiers from the local garrison, the Héraultais-raised 100th Regiment of Infantry, were observed by police agents to join in the cheering.

The ultimatum passed. There was no move from Paris. The tax strike was proclaimed two days later in the flamingly radical newspaper *Le Petit méridional* – with a demand that "every municipal authority, without exception, should within the next three days notify by letter to the Prefect their collective resignation."

The next morning and the next, black flags supplanted the tri-color flying from rebel *mairies*. By 13 June a third of the municipalities in the Hérault had walked out, a quarter in the Aude. Within a few days, half of the elected local authorities in four southern departments had declared for the small-town rebellion. Albert and his core followers on the Argeliers committee were openly talking of declaring a regional parliament. The royalist newspaper *L'Éclair* of Montpellier began stirring up trouble. This was not about sugar, this was secession.

Georges Clemenceau, president of the council and minister of the interior, an old radical with a tough line on enemies within, appealed directly to the delinquent mayors, without success. In Paris, the National Assembly meanwhile had had time to consider their own commission's inquiry. They voted on 17 June for a tax of forty-five francs per quintal on sugar. It was too little too late.

The day before, warrants for the arrest of Marcelin Albert and his followers had been issued, and mandates signed by Clemenceau for troops to be sent into the towns and countryside of the mutinous south. The troublesome 100th Infantry Regiment, made up of conscript farmboys who might feel sympathetic to the *vignerons'* revolt, was urgently posted away from Narbonne to a camp at Larzac in the Cévennes hills.

Albert went into hiding in a bell-tower in his hometown of Argeliers. Dr. Ferroul's barricaded house in Narbonne was surrounded by troopers of the 10th Cuirassiers, brought in by train from Lyon. After a few hours' stand-off, the rebellious doctor surrendered with dignity, pleading for calm. His supporters were incensed, stones flew, carbines were shouldered. On the evening of 19 June Narbonne's sub-prefecture was invested and some of its police defenders were injured. The newly arrived military commander ordered the town square cleared. The breastplated cavalrymen charged. One demonstrator was killed outright, nine injured by gunshot wounds, dozens more by saber slashes.

The upheavals in Narbonne continued all night and through the next day. The following afternoon troops of the 139th Infantry Regiment opened fire on a crowd advancing on the Hôtel de Ville, killing five and wounding ten. There were violent scuffles in Montpellier, and at Perpignan the prefecture was burned down (it seems to have been a royalist stunt). The 17th Line Regiment garrisoning Béziers looked as unreliable as their Montpellier comrades. Made up of two battalions of local conscripts and around a hundred reservists, on the night of 18–19 June the infantry unit was ordered to Agde, a few kilometers away on the Mediterranean coast. When news reached them of the shootings in Narbonne, the reservists mutinied, set the conscripts and their officers to flight, seized the arsenal and distributed weapons and ammunition to a growing band of civilians before heading back to Béziers for a showdown. The paramilitary throng of around five hundred held the vine-growing town for two days before the soldiers agreed to go back to barracks with the promise of an amnesty (they were sent, awaiting trial for mutiny, to a punishment battalion in Tunisia).

Marcelin Albert came down from his bell-tower. In an eventful journey by motor-car and train he managed to reach Paris to plead the *vignerons*' case with Clemenceau himself, who had just won a vote of confidence in the Assembly. They met in his study on a Sunday morning. With all the dignity of his office, with all the cunning of a political street-fighter, Clemenceau ran rings around the café-owner. The vine-growers' case might be just, but their actions had traduced the honor and integrity of France. Persuade the mayors to go back to their duties, preach calm to your followers, the veteran politician argued, and the troops would be withdrawn. He slipped Albert a hundred-franc note "to pay for his return journey."

The revolt of the *vignerons* ended in utter débâcle. Clemenceau briefed journalists that Albert had accepted money to call off the rebellion. The apostle of pure wine returned home to find his own purity besmirched. His followers turned their backs on him. Broken and humiliated, he gave himself up to arrest at Montpellier on 26 June. The revolt fizzled out.

There was a kind of victory. The law of 29 June 1907, passed in the National Assembly nine days after the Narbonne shootings, made sales of sugar in quantities over twenty-five kilograms notifiable to the Ministry, and banned the sale of the most notorious adulterating chemicals. An "anti-fraud service" was established to sniff out the sugarers' more devious activities, and a decree of 3 September officially defined anything sold as wine as "the exclusive product of alcoholic fermentation of the fresh grape or the juice of the fresh grape."

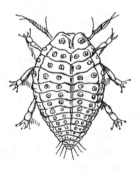

34

The devastator had been beaten. There remained the matter of the prize. What had really saved France's vineyards? It was not the palliative of chemical insecticide. It was not the discovery that the roots of certain American vines were unpalatable to the imported insect – the wine they made was unpalatable to every true Frenchman. It was grafting that had proved the magic bullet. Who had proposed it first?

Léo Laliman of La Touratte, Bordeaux, thought it was him. His case seemed excellent. He had written volubly about the strange new idea during the *oïdium* crisis. He had shown off his American vines to the Beaune Viticultural Conference of 1869. Gaston Bazille, who had begun grafting experiments soon afterwards, was courteous enough in his subsequent pronouncements to defer to the fact that Laliman was the first to advance the idea. Planchon for one also thought he deserved it. Along the way Laliman petitioned the Ministry of Agriculture, the Superior Commission and the Agricultural Society of the Gironde to take up his case. Without success.

The trouble was, as the Duchesse de Fitz-James wrote, "This innovator, the hero and at the same time the villain of the story, was so proud of having imported the American vines that he could never accuse himself of having paid the travel expenses of the phylloxera that travelled with them on the roots."

Officialdom was deliberately ignoring his claims, Laliman alleged in 1879. He had been surrounded with a *cordon*

Left Professor Jean-Baptiste Dumas, permanent secretary of the French Academy of Sciences and chairman from 1870 of the government-appointed Central (later "Superior") Commission on the Phylloxera. Himself a distinguished chemist, Dumas sternly advocated chemical insecticide as the only defense against the aphid. Only after his death in 1885 was grafting formally recognized as an alternative route to redemption for France's vines.

Above Alexis Millardet, the Bordeaux botanist and pioneer geneticist who crucially established which American vine species were truly resistant to the phylloxera and which were compromised.

Left Professor Maxime Cornu of the Paris Museum of Natural History, secretary of the Phylloxera Commission, who conducted important early research on just how the parasitic aphid attacked vine roots and on potential insecticides. Although a life-scientist, he was a powerful ally of the *sulfuriste* camp – switching in the early 1880s to become an advocate of grafting.

Pierre Viala, professor of viticulture at Montpellier, was sent on a government mission to Texas in 1887 to find phylloxera-resistant vines that would grow in chalky soil. The results proved highly contentious.

The stern bust of Camille Saintpierre, pioneer in the fight against the aphid and later director of the Montpellier School of Agriculture, stares out over La Gaillarde, the "university of phylloxera."

The memorial to Jules Planchon, in the Place Planchon opposite Montpellier's railway station. It is inscribed: "The American vine rescued the French vine to triumph over the phylloxera."

The sensuous memorial to Gustave Foëx at La Gaillarde. An allegorical nubile young *américaine* embraces a maturer female form representing France's embattled vines.

The experimental vines at La Gaillarde grow just where they did over a century ago – flanked now by ultra-modern research facilities and laboratories as France educates a new generation of vine-growers to compete in a global market.

sanitaire of lofty disdain, while he was "accused of being the Attila of our vines – thus suffering a martyrdom of which I shall one day write an incredible history." He published the promised work in 1889, *Notice chronologique sur l'origine des vignes américaines résistant au phylloxéra*, but it was more wheedling than sensational, a tiresome résumé of how he had been first with every key observation and innovative suggestion, how Planchon, Lichtenstein and Riley had hogged the plaudits that should have been showered on him, how this self-congratulating "triumvirate" had been responsible for the disasters of the early 1870s when "seven million" non-resistant Concord vines had been planted in the Midi against his own warnings. According to the curious pamphlet, a mysterious "M. Osiris" was now offering a hundred-thousand-franc prize for a practical phylloxera cure to be demonstrated at the centennial exposition then taking place in Paris. It should by rights be his (that prize too went unawarded).

"All of our disciples have been recompensed or given honors, except the hermit of La Touratte," Laliman declared, "who is the enemy of that mutual admiration society whose sole mission has been to exploit the man whose work they do not understand – the lonely vine-grower of the Gironde who has been thrown to the side, who has never received any encouragement even verbally – but whose success has been recognized by foreign authorities and by prize juries prepared to stand up to certain hostile parties . . ." He was still insisting that root-living phylloxera were not – and never had been – an American import.

Laliman was still making supplications in 1897, at the age of eighty. At the session of the General Council of the Gironde on 1 May that year, his advocate, a M. Delbay, declared: "All he asks of France is that she fulfils the obligations she undertook to pay the letter of credit now due – the prize of 300,000 francs. Today, when our vines have been saved from disaster by the methods put forward by M. Laliman

. . . you propose to bargain with an old man who has sunk a large part of his fortune in experiments and hard work which have resulted in the saving of France's vineyards."

The Council concluded a month later: "M. Laliman's claim is rejected because all M. Laliman's work, important though it is, had as its object the substitution of American vines or American roots for our French varieties on their own roots. At no time did M. Laliman try to destroy the insect; nor did he try to any greater extent to stop it doing damage . . ." The president of the council added dispassionately: "In fact the dossier can be withdrawn because the petitioner no longer exists." The Attila of France's vines had died on 3 June 1897. His son kept on trying to obtain, in his words, "The national reward earned by my father, the tireless pioneer who from the moment of the phylloxera's invasion of our vineyards . . . has so greatly contributed to their reconstitution and health." He was no more successful. *Le Journal d'agriculture pratique* gave him a brisk but forgiving obituary: "M. Laliman was the first to suspect the part that American vines might play in the fight against the phylloxera. He was thus one of the first partisans of reconstitution and for this his name will never be forgotten."

The flow of suggestions from other prize-hopefuls to the Ministry of Agriculture had meanwhile slowed to a trickle. The prize sub-committee of the Superior Commission wound itself up with little ceremony in 1900. The glittering *concours* of 300,000 gold francs sat unawarded in the vaults of the Banque de France. It has done so ever since.

Others in the phylloxera struggle found their rewards in academic distinction and the plaudits of a grateful nation. Jules-Émile Planchon had died suddenly on 1 April 1888. With funds raised from a nationwide subscription, a bronze bust was raised on a plinth opposite Montpellier's railway station six years later.* The monument was inscribed: *La vigne américaine a fait revivre la vigne française et la triomphe du phylloxéra – À Planchon, les viticulteurs reconnaissants.* (The American vine made the French vine live again and triumphed over the phylloxera – to Planchon, wine-growers shall remember). It was flanked by a representation of a handsome young *vigneron* offering the professor a bunch of grapes.† The "robust peasant" was poignantly modelled on the painter son of Senator Gaston Bazille, Frédéric, *mort pour la patrie* in the frontier battles of 1870.

M. Vigier, minister of agriculture, performed the unveiling. His eulogy was both solemn and political. The professor's life-story was an object lesson in "rural democracy" – this son of the people, a *montagnard* from the Cévennes, child of a small artisan who left his village to become apprenticed to a pharmacist, had reached the heights of scientific endeavor through hard work and intelligence. How fortunate for France that he had, having overcome years of "unjustified attacks and systematic indifference," been able to proclaim the truth. Phylloxera was the cause, and American vines the solution. Everyone now knew he was right. He had not been beaten. His "apostle's temperament" drove him on. He could both theorize and popularize. Jules-Émile Planchon lived by Arago's dictum for the scientist: "Discover, understand, communicate." Most important, perhaps, he had shown himself to be a master of what the republican hero Léon Gambetta called "the politics of results."

* Announcing the subscription to pay Planchon's family a pension and raise a suitable monument, Armand Lalande famously calculated the direct financial loss to France caused by the phylloxera at ten billion francs, or double the reparations paid to Germany in 1871.
† The Planchon bust was removed by the Germans and melted down for munitions after the takeover of Vichy France in 1942. It was restored in stone in the 1950s.

After the little fête, the minister proceeded to be guest of honor at a splendid banquet at the Montpellier School of Agriculture. Grafted wines were served. The Swiss-born Gustave Foëx wrote a classic manual of viticulture which sold around the world. He ended his career as France's inspector general of viticulture. After his death in 1906 he too was commemorated in allegorical statuary, in the gardens of the Montpellier School of Agriculture, with a naked young *américaine* tenderly embracing a maturer female form (the ancient vine of France).

Alexis Millardet, the visionary of hybrid rootstocks and discoverer of "Bordeaux mixture," acquired a vineyard in the Charente. His *eaux-de-vie* were judged to be "excellent." Twelve years after his death in 1902, a memorial monument was raised in Bordeaux by the Gironde Agriculture Society.

Pierre Viala continued to traverse France preaching the message of the grafted vine. In 1901 he was appointed professor of the National Agronomic Institute of Paris. In 1909 he published, with the energetic Beaujolais *viticulteur* Victor Vermorel, a magisterial seven-volume ampelography. As the process of "reconstitution" spread across Europe, so did Viala's fame. He was heaped with academic honors not just by France but by Italy, Spain, Serbia and Bulgaria. He was working until the end of his life on the condition known as *court-noué* (fan-leaf virus) and on fungal afflictions of vines in Palestine. He died in 1935.

In 1888 the French government sent a delegation to Texas to confer on Viala's host and guide, Thomas Volney Munson, the Chevalier du Mérite Agricole of the Légion d'Honneur. He died in 1913. In 1988 the T.V. Munson Viticulture and Enology Center was opened in his home town, Denison. A statue was raised in Denison's twin town, Cognac.

Charles Riley's career prospered. In 1878 he was appointed entomologist to the U.S. Department of Agriculture in Washington. In 1888 he helped save California's citrus industry by introducing the predatory Australian vedalia beetle to combat scale insects. Riley was curator of the National Collection of Insects at the Smithsonian Museum of Natural History when he died aged fifty-one in 1895 from injuries sustained in a bicycling accident. Upon the death in 1978 of his last surviving child, Dr. Cathryn Vedalia (named for the benign Australian beetle) Riley, a trust was established to found the Charles Valentine Riley Memorial Foundation.

No statue was ever raised to the aphid. The memorial to its far-flung campaigns is in the vineyards of most of the wine-making world – with scientifically managed, globally marketed, varietally branded, marching rows of vinifera grafted onto hybrid roots. Biology defeated the insect; toil, sweat and tears put the vineyards back together again. Towards the end of the second millennium, however, it seemed that biology had another trick to play. It began in California.

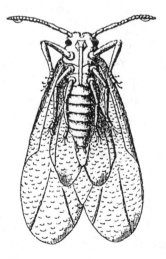

35

In the last quarter of the nineteenth century the phylloxera's twenty-year march through the vinifera-planted vineyards of California had been as implacable as they had been in the Old World vineyards from which European vines had originally been imported decades earlier. Not until the 1890s did farmers begin to adopt the apocalyptic French solution – rip up the dying vines and start again. In 1896 a scientist from the University of California at Berkeley went to France to report on the latest rootstock research. He returned with a variety of an American vine – called "Rupestris du Lot" by French nurserymen, known as "St. George" in California. In spite of its low yields it became the industry standard as a graft-bearer. Slowly the state's wineries began to prosper anew – until Prohibition snuffed them out in 1919. A few survivors sent table grapes by rail to the east – with warnings attached to the crates that they would produce alcohol if inadvertently allowed to ferment.

The long sleep ended in the 1930s. California's reborn wine industry would be informed by scientific principle. The University of California's "model farm" established in 1903 at Davis, twenty miles west of Sacramento, expanded into a separate campus with its own College of Agriculture. Its Division of Viticulture and Enology would become a world leader. In the 1950s its eager graduates began poking into every cranny of Californian vine-growing and wine-making, bringing science to what was still largely a quaint backwoods operation.

One key recommendation was the substitution of St. George

rootstock with the century-old creation of Victor Ganzin – AxR1 – after viticulturalists from Davis concluded that it produced "excellent yields and quality of most wine grapes in most north coastal climates and soils." In spite of its half-vinifera parentage, the hybrid rootstock's resistance to phylloxera was judged more than adequate.

California's vineyards expanded in the 1960s at the same rate as the reputation of their wines. Most of the new planting was on AxR1 rootstock, typically in artificially irrigated, cool, deep soil that had never been used before for vine production. It performed beautifully. There were warnings from the Old World – *viticulteurs* have long memories. On a visit to Napa Valley in 1980 Dr. Pierre Galet, chairman of the Department of Viticulture at Montpellier, pointed out that Ganzin's hybrid had long ago shown its susceptibility in Europe after about fifteen years in the ground. Still, at that time in California there seemed to be no phylloxera depredations. Wherever there are vines, however, the underground aphid is always in malign attendance. Growers happily continued planting on the Aramon-rupestris hybrid – more than twenty-five thousand acres in Napa and Sonoma between 1982 and 1990.

But something was already going wrong. In 1977, in a large vineyard east of the Napa River near Rutherford, a number of vines displayed foliage bunched like "cabbage-heads." Phylloxera was found on the roots of ungrafted vinifera, planted for speed and financial expediency. It was no surprise that they should fail. But the outbreak six years later at St. Helena was on supposedly resistant AxR1.

The Davis scientist Jeffrey Granett began experimenting not with vines but with the insect itself. In 1983 his laboratory published research showing that phylloxera collected from infested AxR1 roots doubled in population size significantly faster than phylloxera collected from other sources. The Davis researchers called the non-damaging populations Biotype A. The fast-growing populations were called Biotype B. Further genetic studies began to indicate that both biotypes were themselves made up of several sub-strains. A Darwinian explanation was offered: "The very rare phylloxera that were able to use these weakly resistant roots for food were able to increase their population while those that cannot feed on them fail to grow and colonize. So, with time, the more virulent insects develop high, damaging populations," according to Professor Granett. The new strain had begun its march. By 1988 twenty Napa and Sonoma County vineyards had reported infections. Several of them had mixtures of rootstocks, so simply attributing losses to the failure of AxR1, with its half-vinifera "blood," was not so straightforward. With new urgency a "Phylloxera Task Force" was formed. In 1989 it at last issued an industry-wide warning that AxR1 was compromised.

There was uproar. This was not some terrible underground unknown that was shrivelling the most pampered vineyards in the world – it was a human blunder. Why had AxR1 been touted in the first place? In 1995 ungrafted vineyards in Santa Barbara County, historically devoid of the pest, began to wither. A year later the voracious Biotype B had been detected in most of Napa County's vineyards and was rapidly spreading in parts of Sonoma County. Wearily, the great wine industry of California realized it must "reconstitute" or perish, just as it had done a century before. The cost of replanting on non-vinifera-parented rootstock was $20,000 an acre. Some owners rang their lawyers; others looked to a new generation of scientists at Davis for redemption.

Big science re-entered the battle that Planchon, Lichtenstein, Cornu, Millardet and Foëx had first joined in the 1870s. The weapons in the new fight would have been beyond their wildest dreams. The spread of infection was tracked by satellite and earth-resources aircraft. DNA technology allowed a fabulous piece of detective work by the University of California scientist Doug Downie, who in 2001 tracked by genetic fingerprinting exactly where the aphid had come from.

"I have now collected grape phylloxera from vineyards in eleven countries around the globe and compared these sequences with those of the populations in the native range," he reported. "Preliminary data suggest most grape phylloxera around the world have originated in the northern United States on the vine species Vitis riparia . . . there is support for the interpretation of at least two independent origins, one from the north central region (Minnesota, Wisconsin, Iowa) into South Africa, and one (possibly two) from the north-east (New York, Ontario) into central Europe. Most insects in California, as well as some in Australia, New Zealand and Peru, seem to have a different source, possibly Vitis vulpina in Virginia."

DNA studies on phylloxera infesting the leaves of wild vines in the south-west United States began to unlock the hugely complex story of how host plant and insect parasite had evolved their biological truce over millennia.*

Experiments funded by the U.S. Department of Agriculture in an ultra-sterile container "the size of a walk-in refrigerator" cultured phylloxera eggs and vine "plantlets" together in a nutrient gel. When the aphid hatched and began to extract nutrient from the roots with varying degrees of success this allowed "new phylloxera-resistance estimates for some forty different plantlets – most grown from samples from the grapevine gene bank at Davis." The experiment took just eight weeks. Planchon, Millardet and Viala spent thirty years trying to do the same, with the whole of France as the field trial.

A century and a half after the discovery that carbon bisulphide killed weevils in French grain silos, the science of insecticides had still failed to find an effective answer. The chemical

---

* In 1998 Doug Downie found that phylloxera on wild native vines in Arizona were incapable of infesting roots, and that there were "sexual forms out in the open on the leaves, probably at all times of day and night . . . Our California phylloxera, to our knowledge, have no sex life at all." Simultaneous Australian research found the same thing. Wherever there is vinifera, the devastator abandons above-ground forms entirely and goes for whatever vine-roots it can find.

sodium tetrathiocarbonate was registered for use in California in 1994. Just as when the aphid reigned in the Médoc, it was found that multiple applications of insecticide could expensively control, but not eradicate, the deep-root-living phylloxera. The insecticide Furadan was tried in the 1980s then banned for use in Napa, Sonoma and Mendocino counties due to its fatal effects on birds. A chemical called cinnamaldehyde was being explored in 1999 after "greenhouse research demonstrated that it may act as an inducer of resistance in susceptible vines."

There was an acute note of alarm in 1999 when researchers in California, Germany and Hungary simultaneously reported that rootstocks with no vinifera parentage in their genome (Vitis berlandieri and V. riparia SO4 and 5C) could host colonies of certain phylloxera biotypes. The scientists qualified the results as being laboratory experiments only, and did not predict a catastrophic failure in the field where these biotypes and supposedly immune rootstocks existed.

But it was a warning. Insecticides were impractical. Mutations in the parthenogenetically reproducing cloned populations of phylloxera had and presumably could again find ways through the defenses of resistant roots. In a bland U.S. Department of Agriculture press release announcing the Davis resistance experiments it was noted:

> *Researchers can examine plantlet roots right after hungry phylloxera attack to see if resistant grapevines form natural chemicals that repel the tiny pests. These compounds may be a key to phylloxera resistance. If so, scientists might be able to trace the chemicals back to the grapevine genes that control them and, after that, perhaps rebuild the genes to boost their effectiveness. Or the scientists might transfer the genes into phylloxera-susceptible vines . . .*

Introducing advantageous genes from one species of the genus Vitis to another to resist parasites and pathogens might seem a logical extension of the work of the nineteenth-century *hybrideurs*. But what about genes from a completely different plant or animal – or one made in a laboratory? Darwin's war of nature might be conducted another way: by denaturing the vine itself.

Large-scale research into the Vitis genome began in the early 1990s. The prime commercial aim was to find new defenses against the vine's biological enemies. The New York State Agricultural Experimental Station, for example, embarked on work to enhance mildew resistance in genetically modified Chardonnay, Pinot and Merlot. By 1996 it could cheerfully announce: "In five to ten years, when transgenic vines and rootstocks become commercially available, the public should have become more accustomed to the consumption and use of transgenic foods. There are already transgenic tomatoes, squash, potatoes and cotton on the market. Improved forms of important grape varieties should not be far behind."

In France the researchers of Moët et Chandon, in conjunction with the French government's Agronomic Research Institute (INRA), began experiments with transgenic Chardonnay in an effort to find a defense against Pierre Viala's *court-noué*.

The University of Florida in conjunction with the U.S. Department of Agriculture had also begun work on vines that might resist Pierce's Disease, an insect-borne bacterial infection that fatally clogged the xylem, or water-conducting vessels, of vines. The disease had proved ruinous in California a century before. This time it was spread by a large, leaf-hopping cicada called the glassy-winged sharpshooter (*Homalodisca coagulata*), native to the south-eastern United States – where it made cultivation of vinifera impossible – which had fatally turned up in California in the early 1990s. By the end of the decade the disease was as big a threat to the state's vineyards as phylloxera.*

In Germany a wine-maker produced a wine from vines genetically modified with a gene from barley – intended to confer resistance to molds. In 2001 the Wine Research Centre at the University of British Columbia announced the creation of a GM yeast that "could eliminate an acid from wine without producing byproducts that cause hangovers."

The bulk of all this effort was aimed at finding means of immunizing rootstocks against various viruses or bacteria for which there was no known chemical treatment. Fungal infections such as the *oïdium*, that gobbled up expensive fungicide and labor, might also be beaten. But what if the old villainess could be defeated? Was there a genetically engineered fix for the aphid?

In 1997 the commercial Dry Creek Laboratories of Modesto, California, announced that they had introduced a gene known as GNA from the snowdrop (*Galanthus nivalis*), already tested in potatoes and tobacco, into the embryos† of three vine rootstocks, Freedom, 101–14, and Teleki, primarily as a defense against virus-carrying nematodes. Fifty plants carrying the gene survived to maturity. Walter Viss, head of the Dry Creek research team, also announced that the genetically modified plants would resist phylloxera attacks – for which, according to the company's statement, there were "no existing methods of control."

---

* The U.S. Patent Office recognized the University of Florida–USDA work in 2001. The experimental Thompson Seedless vines were "inoculated" against Pierce's Disease with laboratory-synthesized genes replicating those found in a wild silk-moth.

† Produced from cultures grown in a laboratory under sterile conditions in an artificial growth medium. The cultures consist of tiny clumps of cells originating from the body of the plant (so-called somatic cells), and not the egg or sperm cells, so that each embryo is a clone of the original plant.

"The grapevine is a compound plant, and we're just working with the rootstock," Viss said. "One advantage of inserting genes only into the rootstock is that they will not find their way into the fruit." Drinkers should not therefore be concerned that wine would start tasting of snowdrop. He promised, however, "to work in future with the scion." The California work might deliver the holy grail itself – an ungrafted, phylloxera-resistant vinifera. Could it really be done?

"It should in theory be possible to isolate the genes from the American species that co-evolved with phylloxera and transfer them to European vines," said Franco Mannini, director of the Turin viticultural center, at a conference organized by Slowine* in 2001. No, it was not possible, said Richard Smart, the famous Australian "flying vine doctor": "It would be very difficult because genes for tolerance or immunity to most diseases and parasites, including phylloxera, that affect vines involve many different chains of chromosomes. This makes the task of introducing tolerance to parasites or disease via genetic modification exceedingly difficult, if not impossible."

Impossible or not, the transgenic anti-phylloxera work continued. The famous Rutherglen research station in Victoria began its own experiments with the aphid-repelling snowdrop gene. The University of Adelaide announced in 1998 its "study of the mechanisms of rootstock resistance . . . This information can be used to breed phylloxera immunity in vine-breeding programs and in the transgenic manipulation of grapevines." Three years later the Australian researchers introduced fluorescence to transgenic Shiraz vines to better analyze the fatal interaction of phylloxera and vinifera roots (the feeding aphids glowed in the dark). To begin with, the scientists were blithely open about their work. Soon they would become much more secretive.

---

* The oeonophile offshoot of the Italian-founded Slow Food, which describes itself as "a movement for the protection of the right to taste."

To the researchers it was straightforward. Science had put the phylloxera back in its box. Genetic engineering might defeat it forever. But wine is not soy. Cultural resistance to all things GM exploded around the world. Wine, the most exquisite expression of human interaction with plants, was the perfect standard around which protesters might rally. Environmental pressure groups like Greenpeace clobbered corporations and retailers. Yes, GM vines might need less pesticide (a good thing), but concerns were expressed that "the genetic diversity of varieties could be lost and that there could be changes in taste, colour and texture of grapes" (a very bad thing).

Industries and governments began to duck for cover. A proposal was made at a 1999 meeting of the International Wine Organisation that genetically modified grape varieties in wines should not be identified as such when it came to the point of sale. It was claimed that identifying them as such would only "confuse the consumer." A meeting of European Union Agricultural Ministers in 2001 sent a directive on the matter back to its technical experts for redrafting – "as its economic impact on the wine industry had not been considered." The same year, a meeting in San Francisco of U.S. wine-trade groups agreed, according to the *New York Times*, "to develop an industry position that, executives say, will support research on genetic engineering, while assuring consumers that no new technology will be used until proved safe."

Wine-drinkers, given suitable labelling, could make their own judgements. It was not just consumers, however, who began to grow concerned. Some wine-makers were by now acutely alarmed. Genetically engineered vines could transform the economics of the worldwide industry as fundamentally as the aphid had done a hundred years before. A conference in London in 2001 was given a stark vision of the future, with global corporations marketing low-cost "perfect" wines as if they were any other beverage, with their varietal flavors determined by transgenic manipulation of fruit and yeasts. Genetically modified materials, it was stated, could get into the wine-making process in indirect ways – in sugar, in gelatine, egg whites (used for fining), even in the cotton screens used as filters.

There was a new revolt of the *vignerons*. In France the ghost of Marcelin Albert stirred when peer pressure and press uproar forced Moët in 1999 to uproot and destroy its transgenic champagne vines. A new champion of *le vin pur* emerged, the highly respectable Philippe Drouhin of Maison Joseph Drouhin of Beaune, who in 1996 founded Terre et Vin de Bourgogne, with a mission to "preserve the high quality of our wines while respecting their typicity and the authenticity of our *terroirs*. To respect our environment and its biological diversity. To ensure our children's future and that of our region."

A Bordelais equivalent soon followed – Terre et Vin de Bordeaux – while in Italy the Slow Food movement and the Associazione Nazionale Città del Vino raised the anti-GM banner. In March 2001, at a conference in Beaune, an umbrella organization was formed, Terre et Vin du Monde. By 2003 almost four hundred French producers, two hundred German and many more in America, Italy and Spain had signed up to its core demand of a ten-year moratorium on research into transgenic vines and the yeasts used to ferment their grapes.

There is a commercial as well as an environmental argument. "Burgundy and Bordeaux," reported the *Wine Spectator* in 2001, "are rife with talk about genetically modified yeast cells [easy to produce because of the simplicity of the microorganism's genome] that can be made to produce specific flavors, which might then compete with *terroir*-driven wines whose aromas are linked specifically to their vineyard sites and microclimate. If these GM vines and yeasts take hold in the Côte d'Or, some fear it will reduce their ability to produce unique Chardonnays and Pinot Noirs – and then, they wonder, what competitive edge will Burgundy have against New World wines?"

Consumer repugnance remains the drag anchor on the advance of transgenic Vitis. In February 2002, however, the European Union issued Council Directive 2002/11/EC, an amendment to an earlier directive on the "marketing of material for the vegetative multiplication of vines." Growers could plant whatever they wished, subject to existing legislation and controls to safeguard human health and the environment.

Research on transgenic vines continues in Europe, the United States, Chile, Israel and Australia. Industry experts declared in 2003 that the production and eventual marketing of GM wine was inevitable.

France's wine-growers had once split between the sulphurists and the Americanizers. Which side would Professors Planchon and Millardet be on in the battle between the *terroiristes* and the genetic engineers? As scientists, they would have embraced the novel; as Frenchmen they might have shared M. Drouhin's concerns. The point is that their solution worked. Grafting beat the aphid and ran the taste of fox to earth. They found the fix within the vine itself. Why mess around with snowdrops and silk-worms? Wine had been saved for the world. Next time a glass is raised, just be grateful for that.

POSTSCRIPT

❧

Professor Lucien Daniel started something with his post-reconstitution grumbles in the 1900s that things were not what they used to be. The debate about pre- and post-phylloxera wines has continued ever since. The philosopher Michel Le Gris got into similar trouble a hundred years later with the publication of his book *Dyonisos crucifié* (1999), in which he protested that science had smothered the veracity of French wines, not just in the cause of beating natural enemies, but in pursuit of profit. The cult of *terroir*, with its fascistic blood-and-soil undertones, was just a cynical front for what had become a food-processing industry, serving up banal products for a global market with infantilized palates. Stabilizers and preservatives added to the wine left detectable undertastes, the Strasbourg oenophile declared. Like Professor Daniel, Le Gris blamed the abandoning of ancient viticultural practices that the long-ago battle against the aphid had seemed to necessitate. The "calamity in the vineyard" had continued ever since:

> *The real disaster is the almost complete disappearance of organic matter . . . Modern cultural methods have left the soil devitalized, populated only by parasites such as the phylloxera . . . for which the means to actually eradicate have never been sought. From that time onwards, new pathogens, new fungi, new viruses, have arrived without*

cease,* all combated of course by chemical means, which only aggravate the imbalance further . . . One wonders how the vineyard was able to be successfully cultivated for five thousand years when one sees the chemical arsenal now considered essential for its survival . . .

There were some Frenchmen who had never given up the old ways, and had been tending the vine and making wine in the pre-phylloxera manner (even if the vines themselves were grafted) for a century or more. Le Gris especially recommended the wines of Château du Puy in the Côtes de Francs *appellation*, not far from Saint-Émilion. Some of the "natural wines" in its cellar were a hundred years old, he told an interviewer, "which goes to prove that, even with very low levels of [preservative] sulphur, wine can be kept for a long time if it is well-made and from vines grown on healthy soil."

Which seems to be freakishly true even of some vintages put in bottles before the years when the aphid reigned. A few still slumber dustily in cellars. Barnacle-encrusted bottles occasionally rise from salvaged shipwrecks. In April 2000 Château Latour brought an 1863 wine, along with several other not quite so venerable vintages, to a set-piece auction at Christie's in New York. At a celebrity preview a sample bottle was opened. It was a Tutankhamun moment. Once it hit the air, would it turn to vinegar in the glass?

The critic of the *Wine Spectator* swooned: "I have tasted pre-phylloxera wines a dozen or so times in my life," he said, "and they never cease to astonish with their purity and clarity on the nose and palate . . . The 1863 Latour had the texture of the finest silk, with aromas and flavors of tobacco, cigar box, leather, berries, plums and wet earth. The flavors remained on

---

* The worldwide spread and prevalence of viral diseases of the vine can be directly blamed on the universality of grafting. Rootstocks may themselves appear symptomless, but in fact be infected.

the palate for minutes after each sip." The experience was worth $4600.

There is a much cheaper alternative. The clever marketeers of Chilean wine expound a simple geographical lesson. The country, flanked on either side by the Andes mountains and the Pacific Ocean, biologically sealed by the arid Atacama desert in the north and icy Patagonia in the south, never admitted the aphid. "As is the case with all of Chile's vineyards, each vine is a non-grafted, original rootstock imported from France during the middle of the nineteenth century before the phylloxera crisis destroyed all of Europe's plants," Chile's prestigious Château los Boldos delights in declaring. As well as in Chile, commercial own-rooted vinifera vineyards have survived in Cyprus, southern Australia, parts of New South Wales and California and much of Russia. Crete stayed immune until 1984. In patches of France and elsewhere, sand still provides an answer. Louis Faucon's solution of winter flooding was largely abandoned in the 1970s.

There are however curious pockets of old vines growing *franc de pied* where they have always grown. Their survival is mysterious, their fermented products almost mythical – haunting wine-lovers' imaginations like oenological coelacanths. In the Champagne, for example, at Ay and Bouzy, a few plots of ungrafted Pinot Noir for some unknown reason did not succumb to the phylloxera's predations of the 1890s. The Bollinger family preserved them and have done so ever since, their workers cultivating the venerable vines manually with traditional tools and by ancient methods. They make Bollinger's unique "Vieilles Vignes Françaises," described as "the living relic of what champagne was a century ago before phylloxera destroyed the vineyards of France."

On a humbler level, in the early 1990s a M. Henri Marrionnet discovered a small plot of ancient Romarantin (an obscure Loire variety) vines growing at Soing-en-Sologne in the Loiret-Cher. His research at Montpellier dated them to the 1850s.

The old vines looked as if "the first gust of wind would blow them down," but Marrionnet got them into production, offering the first vintage for sale in 1998 under the name of "Provignage." Reviews were ecstatic – "Incredible, its taste is like nothing else known," said a satisfied customer, "but one is a little lost without any base of reference to compare it to."

In the Bordelais in 2003, the proprietors of Château Haut-Bailly were making a small amount of wine from "three very old pre-phylloxera vines." "The wine is fine," said its makers, "recently brought into production by technical means, to honor the ancestral style of classic Bordeaux."

The Domaine Charles Joguet in the Val de Loire offers a red Chinon called Les Varennes du Grand Clos, produced from ungrafted vinifera vines. "The results are extraordinary," its makers claim. "Vines grown *franc de pied* always show a weaker degree [of alcohol], . . . but the experience in the mouth is unforgettable, with a sweet and sophisticated texture, rich in tannins and extraordinary silkiness . . ."

In Australia the Tahbilk winery in Victoria was producing a "Reserve Shiraz" from a miraculously surviving ungrafted, pre-phylloxera vineyard said to be the world's oldest plantation of the variety. For the truly adventurous, mysterious pre-phylloxera wines were being offered on smudgy Romanian internet sites.

By the beginning of the third millennium even the most eccentric nostalgists were being catered to. In certain areas of France the ignoble American *producteurs directes* and their hybrids had sustained generations of wine-drinkers. Although new plantings of Noah, Othello, Isabella, Jacquez, Clinton and Herbemont had been banned by government decree in 1934 (Franco-American hybrids were reprieved), the foxy vines lingered on in backwoods uplands until the 1960s. Production of their wine remained legal for *consommation familiale*.

In 1993 a group of enthusiasts in the Ardèche began to reclaim an old American vineyard from the brambles in the high valley of Beaume. They formed a society called La Mémoire de la Vigne to commemorate the years when for a peasant *vigneron* it was American wine or nothing. Some drinkers had become quite attached to "the wine of the resistance, the wine of the anarchists, the wine that drives you mad." It was hoped the phylloxera-era vineyard might become a tourist attraction, and it began to produce something which, as its makers had to admit, had "a slightly puzzling rural taste." Metropolitan sophisticates, however, loved it.

At a blind tasting of Jacquez and Herbemont, the wine expert Stéphane Debaille, a former *sommelier* of the year, declared them to be: "Of consistent color, young, fresh, pleasant yet well developed . . . their structure based on notes of jammy black fruits with touches of sweet spices – a complex and subtly nuanced nose, the curvature of attack finished by a velvety completeness – acidity, alcohol and tannins in perfect harmony . . ."

What a shame such opinions had not been expressed at the Montpellier Wine Congress of 1874, or in Louis Pasteur's laboratory. Much unnecessary effort might have been spared.

ACKNOWLEDGEMENTS

I wish to thank the staff of the British Library (St. Pancras and Colindale); the entomology library of the Natural History Museum, South Kensington; the Royal Botanic Gardens, Kew; Down House, Kent; l'Office de Tourisme, Roquemaure; la Bibliothèque Centrale, École Nationale Supérieure Agronomique-INRA, Montpellier; and the Shields Library, University of California, Davis. Special thanks are due to Kate Pickard, Mme. Dominique Fornier, Mme. Christian Rouse and M. Marc Mascré, Mlle. Mélanie Boissel, Daryl Morrison, John L. Skarstad, Axel Borg – and Mark L. Parsons of the Robert Mondavi Winery, Oakville, California, for being such a welcoming host.

I am grateful to the Royal Horticultural Society, Wisley, for the supply of vines, Messrs. Oddbins of Bellevue Parade, Wandsworth, and Messrs. El Vino of New Bridge Street, Blackfriars, for provision of further essential research material, to my daughters Katy and Maria for assistance on a memorable transit of California – and I must thank my wife Clare and son Joseph for putting up with a kitchen table covered for months in pictures of disagreeable insects.

273

CHRONOLOGY

1798   J.-J. Dufour with followers from Switzerland establishes "First Vineyard" in Kentucky. Two years later their European vines have perished

1815   Atlantic trade re-established following the Napoleonic wars. Small-scale importation of American vines to Europe begins

1826   J.-L. Berlandier embarks on botanical mission to Mexico

1831   James Busby successfully ships living vines in moss-packed parcels from Montpellier (via Kew) to New South Wales

1835   Nathaniel Ward develops the "Wardian Case" terrarium, allowing the transoceanic transport of living plants

1838   First purely steam-powered crossing of the Atlantic (in fifteen days)

1840s   Nicholas Longworth establishes large-scale wine production around Cincinnati, Ohio, with native vines

1845   "Late blight," an American-originating fungal infection of the potato, hits crops in northern Europe. Mass starvation in Ireland. Comte Odart publishes *Ampélographie Universelle*. The *oïdium*, a parasitic fungus of the vine originating in America, is found in a greenhouse in Margate, Kent; from there it spreads through southern France, Italy and Spain

1855 onwards   Spraying foliage with powdered sulphur becomes a generalized effective method of defeating the *oïdium*

1856	U.S. botanist Asa Fitch publishes notice in New York of an aphid (he calls it Pemphigus vitifolii) that lives in leaf galls on vines native to eastern North America
1859	Charles Darwin publishes *On the Origin of Species by Means of Natural Selection*
1860	Léo Laliman of Bordeaux publishes a book in praise of American vines. He advocates grafting as an anti-*oïdium* measure
1860 onwards	Generalized import of American vines into Europe as a measure against the *oïdium*. French vine-growing regions linked to Paris by rail
1862	M. Borty, wine merchant of Roquemaure, Gard, receives a case of vines from New York and plants them in his walled vineyard
1863	Leaves of vines in a Hammersmith (west London) greenhouse develop strange blisters. The phenomenon is brought to the attention of J.O. Westwood, "Insect Referee" of the *Gardener's Chronicle*, but he does not comment in print
1864– 65	Vines around Roquemaure mysteriously begin to wither and die
1866	Isidor Bush establishes "Bushberg" vine nursery in Missouri. C.V. Riley publishes notice of a "grape-leaf gall-louse" in *Prairie Farmer* newspaper of Chicago
1867	Greenhouse vines at Powerscourt, Ireland, show leaf blisters, then sicken and die. Unknown insects are found attacking their roots. A similar infestation is reported in Cheshire, England. Exposition Universelle in Paris. The U.S. government sends a special commission to report on the French wine industry. "Foxy" American wines are awarded zero in blind tastings. Infection in lower Rhône valley spreads remorselessly. First press reports of a new disease of the vine
1868	Spring: The Hérault Agricultural Society forms "La

Commission pour Combattre la Nouvelle Maladie de la Vigne." 15 July: Félix Sahut and J.-E. Planchon find swarms of unknown insects on the exhumed roots of vines near Saint-Rémy in the Bouches-du-Rhône. 3 August: Planchon announces "new insect" to the Academy of Sciences as Rhizaphis vastatrix – the "root aphid devastator." The Paris entomologist Professor Victor Signoret suggests it be called Phylloxera vastatrix, the "dry-leaf devastator." 28 August: Planchon observes a captive root-living phylloxera moult into a winged form. Darwin publishes *The Variation of Animals and Plants under Domestication*. In a paper to the Ashmolean Society of Oxford Professor Westwood names the vine-leaf insect Peritymbia vitisana

1869   30 January: Westwood publishes details of the earlier Hammersmith and Cheshire vine infestations in the *Gardener's Chronicle*. 11 July: Planchon and brother-in-law J. Lichtenstein find leaf-galls on a "Tinto" vine at Sorgues in the Vaucluse. The insects within seem the same as those already found on roots. Soon afterwards a root-attacking infestation is reported near Bordeaux. The chemist Baron Thénard tries to kill the underground insects with carbon bisulphide. The infected vines die. 5 September: Lichtenstein publishes theory that the aphid had somehow come from America. Riley sends samples of Missouri leaf-gall phylloxera to Professor Signoret in Paris. He concludes that they are the same as those now attacking vine-roots in the Midi but insists the infestation is a symptom, not the cause, of the malady. November: Laliman presents his American vines to the Beaune Wine Congress. He and Gaston Bazille of Montpellier begin grafting experiments soon afterwards

1870   July: French Imperial Ministry of Agriculture and Commerce appoints a committee on the phylloxera and offers a 30,000-franc prize for a practical remedy. 16 July: France declares war on Prussia. 15 September: Siege of Paris begins. Signoret communicates the results of continuing experiments by balloon post. December: Riley finds root-living phylloxera on vines at Bushberg, Missouri

1871   July: Riley arrives in France to consult with Signoret. He joins Planchon and Lichtenstein in field investigation in the Midi. A few winged forms are found but no male can be identified

1872   Attempts to halt the plague's progress of uprooting and burning fail. It becomes clear to Planchon that the infestation is spreading of above-ground and underground migration of wingless "crawlers." July: Experimental vineyard to test supposed remedies set up at Montpellier School of Agriculture. September: At the Lyon Agricultural Congress Louis Pasteur proposes finding a biological enemy of the phylloxera. Riley sends samples of an apparent native American predator to Planchon. Outbreaks in Douro valley in Portugal, Austria and the Crimea

1872–73 Mass import of vines from Missouri into the Midi

1873   August: Planchon leaves on government-funded fact-finding tour of America. Outbreak reported in California. Professor Georges Balbiani identifies male and female forms of an aphid that infests oak trees

1874   July: French government prize for a practical cure increased to 300,000 francs. Prefects given discretionary powers to ban imports of alien vines. August: Male phylloxera identified. October: The Montpellier Wine Congress jury samples first fruits from locally grown American vines. The results are grim. First outbreaks in Spain and Germany

1875   Planchon publishes *Les Vignes américaines, leur culture, leur résistance au phylloxéra et leur avenir en Europe*. Grafting experiments in the Midi and Bordelais begin to show promise. First laboratory observations of male and female phylloxera mating – and the resulting "winter egg." Balbiani declares that by destroying the egg, the plague can be arrested. P-L-M railway company establishes an anti-phylloxera service to promote use of carbon bisulphide insecticide on an industrial scale. First outbreaks in Hungary and Australia (at Geelong, near Melbourne)

1876   March: Prefects in vine-growing departments ordered to form "study and vigilance committees." Professor Alexis Millardet of Bordeaux begins research on hybrid Franco-American rootstocks

1877   August: Congrès Phylloxérique Internationale held in Lausanne, Switzerland. October: Republican majority in French National Assembly

1878   Agreement of Berne sets international rules on notification of outbreaks and restriction of movement of plant materials across frontiers. 15 July: Law passed banning import of alien vines other than in districts already declared "phylloxerated." The Superior Commission on the Phylloxera is formally established with executive powers. Troops enforce insecticidal treatment in Burgundy

1879   2 August: Second law closes legal loopholes. Local "anti-phylloxera syndicates" are granted state subsidies for chemical insecticide. Senate rejects tax relief for American vines. First outbreak in Italy

1879 onwards   State-licensed nurseries established in the Midi to mass-produce rootstock and grafted vines

1880   Joseph Audibert of Marseille publishes *The Art of Making Wine with Dry Raisins*. The production of such brews (and others made from sugar and dye)

	becomes generalized. Outbreaks in Serbia and Slovenia
1881	Phylloxera Congress of Bordeaux results in victory for the *américainistes* over the *sulfuristes*. The jury however refuses to pronounce on wines from grafted vines
1884	Superior Commission admits for the first time the advantages of American vines over insecticide
1885	Outbreak in Algeria
1886	First outbreak in South Africa
1887	Pierre Viala embarks on mission to Texas to find resistant vines that will grow in chalky soil. T.V. Munson of Denison recommends the "mountain grape" (Vitis berlandierii). December: Law passed in French Assembly giving four-year tax break on land newly planted or replanted with American vines
1888	1 April: Death of Planchon
1888– 89	Large-scale importation of Berlandieri vines into the Cognac region. They prove almost impossible to propagate
1889	Law passed defining "wine" as the result of the fermentation of fresh grapes
1890	The phylloxera reaches the Champagne. Outbreak in New Zealand
1890 onwards	Large-scale reconstitution using grafted vines begins in Burgundy and the Bordelais
1893	For the first time in fifteen years France's national wine harvest exceeds pre-phylloxera levels – fifty million hectolitres
1895	Death of Riley
1896	California begins reconstruction with grafted vines
1900	Virtually the whole of France declared phylloxerated

1901   Professor Lucien Daniel of Rennes declares reconsti-
            tution with grafted vines to have been "a disaster."
            French wine production meanwhile outstrips na-
            tional consumption, and the price collapses
1903–4   "Fraudulent claret" scandals rock world wine-trade
1905–7   "Revolt of the vignerons" in the Midi against al-
            leged fraudsters and sugarers. It is violently sup-
            pressed by military force
1960   Large-scale planting in California on AxR1 root-
onwards   stock – as recommended by scientists at UC, Davis
1983   Vineyard at St. Helena, California, afflicted by root-
            sucking phylloxera. Scientists identify a new "bio-
            type" of the aphid
1989   AxR1 declared compromised. Large-scale phyllox-
            era predation of vines in Napa Valley and Sonoma,
            California
1992   Beginning of research on Vitis genome and
onwards   transgenic vines
1997   Research begins in U.S. and Australia on "phyllox-
            era-resistant" GM vines
2002–3   Small-scale outbreaks reported in New Zealand,
            Australia and California

SOURCES

Page

xix THE INSECT PLAGUE *San Francisco Chronicle*, 13 July 1992

xxi THERE IS NO WAY *Grapevine Remote Sensing Analysis of Phylloxera Early Stress* (University of California, Davis, 1996)

4 WHEN AMERICANS Wilder, "Report of the United States Commissioners [to the Paris Universal Exposition, 1867] . . . ," p.19

4 IT WAS TO REMEDY ibid., p.5

5 RELIANCE CANNOT ibid.

5 IT IS NOT UNUSUAL ibid.

5 ONE OF THEM BORE Flagg, *Three Seasons . . .*, p.36

6 THE MORE OF THE Wilder, p.13

6 IT WAS RATHER TOO Flagg, *Three Seasons . . .*, p.137

7 DROWNED IN THE Nicholas Longworth, "On the Culture of the Grape and the Manufacture of Wine" (Cincinnati, 1847), in *Report of the U.S. Commissioner of Patents for the Year 1847* (Washington, 1848), p.465

7 PISSAT DE RENARD Garrier, *Histoire sociale . . .*, p.219

8 A FRENCH WINE-GROWER Wilder, p.13

8 AMERICAN CONSUMERS ibid., p.5

8 THERE SEEMS TO BE NO ibid., p.7

8 VITICULTURE IN FRANCE Guyot, "La Viticulture et ses produits," p.602

10 THE MOST NOBLE Dunn, *The Horticultural Handbook*, p.166

11 I SAW CLEARLY M. Dunn, "On the Extirpation of the Vine Pest," in *Journal of the Horticultural Society of London*, 1872, Vol. 3, p.81

12 THANKS FOR THE VINE-LEAVES ibid.

14 LEAVES FROM A J.O. Westwood, "New Vine Diseases," in *The Gardener's Chronicle and Agricultural Gazette*, 30 January 1869, p.109

14 A VINE LEAF COVERED ibid.

15 IN THE LATTER ibid.

16 A DAISY FROM A C. Darwin to J.D. Hooker, September 1846, in F. Darwin (ed.), *The Life and Letters of Charles Darwin* (New York, 1905), p.314

16 PLANTS ARE SPLENDID Darwin to Huxley, 11 May 1880, in F. Darwin (ed.), *More Letters of Charles Darwin*, Vol. 1 (London, 1903), p.307

17 A YOUNG MUSCAT Darwin,

The Movements and Habits of Climbing Plants, p.138

17 AS STRIKING AN ibid., p.143

17 INTENTION WAS TO Browne, *Charles Darwin*, p.170

18 STRUGGLE FOR LIFE Charles Darwin, *On the Origin of Species* (1st edition, London, 1859), title page

18 SINCE REMOTEST Darwin, *The Variation of Animals and Plants under Domestication*, p.352

18 THE BEST AUTHORITIES ibid.

18 THE CULTIVATED VARIETIES ibid.

22 DELICIOUS SUMMER Savage, *André and François André Michaux*, p.241

22 THE BEST VINE ibid., p.238

22 ABOUT SIXTY YEARS J. Le Conte, "American Grape-Vines of the Atlantic States," in *Report of the U.S. Commissioner of Patents for the Year 1857* (Washington, 1858), p.225

23 THE SUBJECT IS Rafinesque, *American Manual of the Grape Vines . . .*, p.5

23 NOT PROSPERED Berlandier, *Journey to Mexico . . .*, p.299

24 ONE SPECIES WITH ibid., p.300

24 THE INHABITANTS OF BÉXAR ibid.

25 THERE ARE SOME WILD Lenoir, *Un Traité de la culture de la vigne . . .*, p.84

25 NAUSEATING Cazalis-Allut, "Sur la greffe de la vigne, 1832," in *Oeuvres agricoles* (Paris, 1865), pp.4–9

25 DELICATE AND PLEASANT ibid.

25 I SHALL BE REDUCED Odart, *Ampélographie*, p.160

27 PROFESSOR DELISLE J. Busby, *Journal of a Recent Visit . . .*, p.86

27 BOXES WERE LINED ibid., pp.91–2

29 DROPSY R.N. Strange, "Resistance of Plants to Disease," www.ucl.ac.uk/biology (London, 2002)

30 A DUSTING OF E.C. Large, *The Advance of the Fungi* (New York, 1940), p.44

31 THE LITTLE VINEYARD *Cozzens' Wine Press* (New York), 20 August 1854, p.17

31 THE DOURO COUNTRY ibid., p.18

32 THE VINE AGAIN ibid.

32 THE GRAPE IS NOT ibid.

34 THE SMALL QUANTITY W.J. Flagg, *Handbook of the Sulfur Cure* (New York, 1870), p.47

34 SUBJECT TO OPHTHALMIA *The Times*, 26 August 1863, p.12

35 THE CURIOUS FACT *The Times*, 24 June 1858, p.12

35 DURING THE VINE DISEASE Darwin, *The Variation . . .*, pp.353–5

36 CONSPIRACY OF SILENCE F. Darwin (ed.), *The Life and Letters of Charles Darwin*, p.539

36 DEFICIENT IN THE BASIC Browne, *Darwin*, p.260

38 ALL LOVERS OF PROGRESS Laliman, *Reformes viticoles*, p.52

38 ATTEMPTS TO CULTIVATE ibid., p.15

41 BORNE IN INDIAN-FILES Weber, *Peasants Into Frenchmen*, p.217

41 ITS USE IN FAMILY Guyot, "La Viticulture et ses produits," p.601

41 TODAY IT IS EUROPE ibid.,
p.600

41 THE VINEYARD OCCUPIES
ibid., p.601

43 FRIEND FROM AMERICA
information from M.
Claude Nova, ancien President of the
Office de Tourisme,
Roquemaure (Gard)

44 TURNED YELLOW Laval, *De
la Maladie de la vigne*, p.3

45 A NEW MALADY *Journal
d'agriculture pratique*, 1867,
Vol. 2, p.39

45 SULPHUR, LIME, COAL TAR
quoted in Planchon and
Saintpierre, *Premières
expériences* . . ., p.15

45 NO DIFFERENCE BETWEEN
YOUNG *Journal d'agriculture
pratique*, 1867, Vol. 2, p.39

46 MOST OF THE VINES
M. Delorme, in *Messager
agricole du Midi*, 1868–9,
5 August, Vol.11, pp.225–8

46 DIGGING ONE UP ibid.

47 AS IF LIFE WOULD *Messager
agricole du Midi*, 1871–2,
Vol. 12, p.184

47 A SHOUT OF DISTRESS
Barral, *La Lutte contre le
phylloxéra*, p.9

48 WANTING AN ACCURATE
Journal de l'Agriculture,
January 1869, p.18

49 VERITABLE CEMETERY Sahut,
Les Vignes américaines

49 THE DELEGATES STUDIED
Revue des deux mondes,
1 February 1874, p.546

50 THE SEVERE COLD *Journal
d'agriculture pratique*, 1868,
Vol. 2, p.281

50 KILL THE INSECTS *Journal
d'agriculture pratique*, 1870,
p.916

50 WE ADMIT THAT THE INSECT
ibid.

51 HEALTHY Sahut, *Un Épisode
rétrospectif* . . ., p.6

51 WE WERE ABLE TO OBSERVE
Sahut, *Les Vignes américaines*

51 UNE LUNE Vialla, *Le
Phylloxéra et la nouvelle
maladie de la vigne*, p.31

51 UNE TACHE D'HUILE ibid.

51 MAKE ITS VOYAGES *Journal
d'agriculture pratique*, 1869,
Vol. 2, p.370

52 ROOT APHID DEVASTATOR
*Comptes rendus de l'Académie
des Sciences*, 3 August 1868,
p.336

52 AGITATION ON ALL SIDES
Planchon and Saintpierre,
Premières expériences . . ., p.4

53 MORE OR LESS NUMB ibid.,
p.28

54 THESE SPIRITED
YOUNGSTERS *Comptes rendus
des travaux du congrès
viticole et séricicole* . . . (Lyon,
1873), p.121

55 WING-BUDS ON TWO SIDES
ibid., p.30

55 PHYLLOXERA QUERCUS Boyer
de Fonscolombe, "Description
du genre phylloxéra," in
*Annales du Société
Entomologique de France*
(Paris, 1834), p.223

55 THIS FORM INDEED
Planchon, "Le Phylloxéra en
Europe et Amérique. I:
l'Origine du phylloxéra, ses
ravages et les moyens de le
combattre," *Revue des deux
mondes*, 1 February 1874,
p.547

56 IN A BOX ONE METER
*Comptes rendus de l'Académie
des Sciences*, 1868, Vol. 67,
p.593

56 IN THE FIRST ibid., p.590
57 HABITUALLY CHARGED WITH
   RAIN ibid., p.594
58 WHY SHOULD WE DECLARE
   quoted in Laliman, "Étude sur
   les divers phylloxéra," p.323
58 IN THE AUTUMN OF 1867
   Westwood, "New Vine
   Diseases," p.109
59 IN FRANCE, WHERE ibid.
60 ON THE FOOTINGS OF SOME
   VINES Journal d'agriculture
   pratique, 1869, Vol. 2, p.404
61 PHTHISIE GALOPANTE Vialla,
   Le Phylloxéra et la nouvelle
   maladie de la vigne, p.25
61 I HAD CULTIVATED quoted in
   Pouget, Histoire de la lutte
   contre le phylloxéra, p.5
62 OF WHICH THREE BRANCHES
   Cornu, Études sur le
   phylloxéra vastatrix, p.13
62 FAT AND TORPID Vialla, Le
   Phylloxéra et la nouvelle
   maladie de la vigne, p.80
62 IN A STATE OF MIGRATION
   ibid.
62 GLOBULAR GALLS Asa Fitch,
   "Annual Report of the New
   York Agricultural Society for
   1856," quoted in Signoret,
   "Phylloxéra vastatrix . . .,"
   p.565
63 IT IS NOT IMPOSSIBLE ibid.,
   p.566
65 I WELL REMEMBER Husmann,
   The Cultivation of the Native
   Grape, p.18
66 ALL AMERICANS TRULY FREE
   ibid., p.25
68 THOUSANDS OF GALLS Riley,
   Third Annual Report . . .,
   p.89
68 ALL THE VINES GROWING L.
   Vialla, "Rapport sur la
   maladie de la vigne," in

Journal d'agriculture pratique,
   1869, Vol. 2, p.601
68 THE APHID WAS
   EVERYWHERE ibid.
69 BAD CULTURE Signoret,
   "Phylloxéra vastatrix . . .,"
   p.569
69 NEW VINE MALADY
   Trimoulet, Rapport sur la
   maladie nouvelle de la vigne,
   pp.4–5
70 SYSTÈME PHYLLOXÉRIQUE
   Trimoulet, 3e Mémoire . . .,
   p.7
70 THE PHYLLOXERA MAY HAVE
   EXISTED Vialla, Le Phylloxéra
   et la nouvelle maladie de la
   vigne, p.45
74 SMALL AS THE ANIMAL Riley,
   Sixth Annual Report . . .,
   p.40
75 AN EGG MOUNTED ON SIX
   LEGS Mayet, Les Insectes de
   la vigne . . ., p.89
77 BY LEAVING THE TROPICS
   Galet, Les Maladies et les
   parasites de la vigne, p.1091
80 TO BE AWARDED TO THE
   ONE "Report addressed to the
   French minister of agriculture
   and commerce appointed to
   inquire into the new disease
   that is afflicting the vine, in
   Papers respecting the
   phylloxera vastatrix or new
   vine scourge" (Melbourne,
   1873), pp.17–18
80 THE MINISTER HAS SHOWN
   ibid., p.16
85 THE BLIGHTING EFFECTS
   Riley, Third Annual
   Report . . ., p.86
85 SINCE THE ABOVE ibid.,
   p.86fn
86 NOT IN HOPE OF Catalogue
   de Bushberg, p.9

86 WITH OUR FULL AUTHORITY ibid., p.10

86 WE THUS SEE Riley, Fourth Report . . ., p.64

86 THE PRINCIPAL REASON Bush, *Les Vignes américaines*, p.10

87 YET LIVING PROGENY Riley, Fourth Report . . ., p.69

87 THE INSECT IS FALSELY W. Saunders, "Report of the Commissioner of Agriculture of the Operations of the Department for the Year 1876" (Washington, D.C., 1877), p.72

88 APHIDAE, I REPEAT Riley, Fifth Report . . ., p.68

88 IN JULY 1871 Planchon, *Les Vignes américaines* . . ., p.62

89 VERY REMARKABLE COINCIDENCE Planchon and Lichtenstein, *Le Phylloxéra en Angleterre et en Irlande*, p.3

89 NOT JUST FROM ibid., p.45

89 HERE WE HAVE AN INSECT Riley, Third Report . . ., p.84

89 UPON THE ROOTS ibid., p.88

90 INSECT CAN BE TRANSPORTED ibid., p.89

91 THE MOST CHARACTERISTIC "Report addressed to the French minister of agriculture and commerce . . ." op. cit., p.13

93 EVIL STAYED INVISIBLE ibid.

93 SOME SAVANTS HAVE Planchon, "La Question du phylloxéra en 1876," *Revue des deux mondes*, 15 January 1877, p.249

94 BRING IT TO THE ATTENTION *Journal d'agriculture pratique*, 1871, Vol. 2, pp.747–8

94 A FEW PARCELS OF VINES Iain Stevenson, "The Diffusion of Disaster: The Phylloxera Outbreak of the Département of the Hérault, 1862–80," in *Journal of Historical Geography*, 6, 1, 1980, p.61

94 ALTHOUGH AN INDEMNITY *Messager agricole du Midi*, 1871–2, Vol. 12, p.184

95 IN THE INTERESTS OF *Messager agricole du Midi*, 1870–1, Vol. 11, p.237

95 DOUBTING THOMASES *Messager agricole du Midi*, 1871–2, Vol. 12, p.185

96 THE CLOSE PROXIMITY Stevenson, "The Diffusion of Disaster," op. cit., p.57

96 THE POPULATION OF *Messager agricole du Midi*, 1871–2,Vol. 12, p.275

96 I AM ASSURED THAT *Messager agricole du Midi*, 1873–4, Vol. 14, p.215

97 BEGIN TO VIBRATE Planchon and Lichtenstein, *Le Phylloxéra (de 1854 à 1873)*, p.20

97 IN CAPTIVITY WE HAVE ibid., p.21

98 IT IS POSSIBLE ibid., p.19

99 THE WEATHER IS MAGNIFICENT *The Times*, 15 August 1872

99 WEAR A SICKLY APPEARANCE Murray to Granville, 3 July 1872, in "Report addressed to the French minister of agriculture and commerce . . .," op. cit., p.3

100 FRENCH OENOLOGUES Crawfurd to Murray, 29 June 1872, ibid., p.4

101 THE INTRODUCTION HERE Hunt to Lyons, 2 July 1872, ibid., p.7

101 HAD NOT NOTABLY EXPANDED ibid., p.9

101 DEAD AND SICK VINES ibid.

101 UNPROVEN RUMORS Dupont to F. Duval, 9 July 1872, ibid., p.11

102 A JUST PUNISHMENT quoted in Garrier, *Vignerons du Beaujolais* . . ., p.123

102 THE SPIRIT OF *Comptes rendus des travaux du congrès viticole et séricicole* . . . (Lyon, 1873), p.124

104 FACILE ARGUMENTS J.-E. Planchon, in *Revue des deux mondes*, 1 February 1874, p.554

104 HOW CAN ONE BELIEVE Planchon and Lichtenstein, *Le Phylloxéra. Résumé pratique et scientifique*, p.14

104 IF A WOLF EATS A SHEEP *Revue des deux mondes*, 1 February 1874, p.557

105 PHTHEIR OR VINE-LOUSE *L'Éclectique*, 24 January 1870, reported in Planchon and Lichtenstein, *Le Phylloxéra. Faits acquis* . . .

105 GABEL *Journal de Lunel*, 28 March 1871, reported in Planchon and Lichtenstein, *Le Phylloxéra. Faits acquis* . . .

105 CAN ONE IMAGINE *Revue des deux mondes*, 1 February 1874, p.547

105 IN OUR REGION Riley, Fifth Report . . ., p.69fn

106 ALWAYS FULL OF EXOTIC PLANTS *Revue des deux mondes*, 1 February 1874, p.551

106 INTELLIGENT AND EDUCATED ibid.

107 MONSIEUR BORTY *Comptes rendus de l'Académie des Sciences*, 1874, Vol. 78, p.1094

107 AMERICAN VINES CULTIVATED *Revue des deux*

mondes, 1 February 1874, p.551

108 NUMEROUS VINE SHOOTS ibid.

108 EARLY VINE INTRODUCTIONS Ordish, *The Great Wine Blight*, p.5

108 STATUTORY INSTRUMENT NO. 1758 The Plant Health (Great Britain) Order, 1987, part II, schedule A, item 10, HMSO

109 MRS S.J. KELLOGG OF CINCINNATI Boehm, *The Phylloxera Fight*, p.11

109 ENTOMOLOGISTE EN TITRE Planchon, *Les Vignes américaines* . . ., title page

109 NO VARIETY OF IT Riley, Third Report . . ., p.89

110 BOTANISTS AND EXPERIENCED Riley, Third Report . . . ibid., p.90

110 WE SEE IN THE GENERAL Bush, *Les Vignes américaines*, p.86

110 IN THE WINTER OF 1870–71 Lichtenstein, *Histoire du phylloxéra*, p.29

112 APPEAL TO MEN OF SCIENCE "Report addressed to the French minister of agriculture and commerce . . .," op. cit., p.16

112 AN AVALANCHE OF *Revue des deux mondes*, 1 February 1874, p.560

113 FANTAISISTES *Revue des deux mondes*, 15 January 1877, p.241

113 GENTLEMEN. THESE ARE *Comptes rendus du congrès viticole de Lyon 1872*, p.134

115 RELUCTANT TO SPEAK ibid., p.140

115 THE SAME THOUGHT ibid., p.142

116 *PESTIVORE AU MOYEN*
Garrier, *Le Phylloxéra: Une
guerre de trente ans*, p.83

116 DRIED HUMAN URINE
Charmet, *Le Phylloxéra*, p.16

117 AT VERY HIGH U.S.
Environmental Protection
Agency. Health and
Environmental Effects Profile
for Carbon Disulfide
(Cincinnati, 1986)

120 ALL MOVEMENT CEASES
P. Mouillefert, *Le Phylloxéra.
Moyens proposés pour le
combattre* (Paris, 1876),
pp.80–1

121 THIS MAN OF PROGRESS P.
Mouillefert, "Excursion dans
les pays phylloxérés," in
Journal d'agriculture pratique,
1876, Vol. 2, p.616

121 PROTESTANTS BY RELIGION
ibid., p.617

124 *PROCÈDE POLLIER Résultats
des divers procédés de
guérison proposés à la
commission pour combattre la
nouvelle maladie de la vigne*
(Montpellier, 1873), p.4; also
*Annales du Ministère de
l'Agriculture*, 1876, pp.190–8

124 EVEN WITH A RELATIVE
Planchon, "La Question
phylloxérique en 1876," in
Revue des deux mondes,
15 January 1877, p.264

124 ALL INSECTICIDES ARE Riley,
Fifth Report . . ., p.71

125 THE REWARD WE BELIEVE
Scientific American,
12 September 1874, p.162

125 SHOULD THE REMEDY
Scientific American,
10 October 1874, p.231

125 MORE THAN SIX HUNDRED
"Rapport de la sous-

commission . . .," in *Journal
d'agriculture pratique*, 1875,
Vol.1, pp.322–3; see also
Trimoulet, *5e Mémoire . . .*,
pp.5–49 for a list of over five
hundred proposed remedies

126 THE PERUSAL OF *Revue des
deux mondes*, 15 January
1877, p.255

126 ILLUSIONS AND DECEPTIONS
ibid., p.252

126 BEATING WHEELBARROW
Foëx, *Cours complet de
viticulture*, p.506

127 IN ALL THE PARTS
Anonymous, *La Vigne . . .*, p.8

130 INSECTICIDAL ACTION Galet,
Les Maladies . . ., p.1304

132 COULD HAVE BEEN
NORMANDY Planchon, *Les
Vignes américaines . . .*, p.19

132 THE ABSENCE OF ibid.

132 PURITANICAL "Notes du
voyage," in Morrow, "The
American Impressions of a
French Botanist," p.72, and
Morrow, *The Phylloxera
Story*

132 A SIMPLE BOY Planchon, *Les
Vignes américaines . . .*, p.22

133 PRINCIPAL QUESTION ibid.,
p.31

133 UNPLEASANT LINGERING
"Notes du voyage," in
Morrow, "The American
Impressions . . .," op. cit.,
p.73

133 TO FIND THE Planchon, *Les
Vignes américaines . . .*, p.34

133 BUT FRENCH IN HIS ibid.

133 AN OCEAN OF VERDURE ibid.,
p.52

134 CROSSING ALL OBSTACLES
ibid.

134 SAD-LOOKING "Notes du
voyage," in Morrow, "The

American Impressions . . .,"
op. cit., p.74
134 VERY AGREEABLE Planchon,
Les Vignes américaines . . .,
p.58
135 ONLY TWO OR THREE "Le
Phylloxéra en Europe et en
Amérique. II: La Vigne et vin
aux États-Unis," Revue des
deux mondes, 15 February
1874, p.917
135 BROUGHT FROM ALL ibid.,
p.931
135 TO JUDGE IN DETAIL ibid.,
p.932
135 THEY ARE SIMPLY Revue des
deux mondes, 15 February
1874, p.934
135 NOTES OF TASTING Planchon,
Les Vignes américaines . . .,
p.78
136 IT IS NOT ibid., p.936
136 BECAUSE OF THE ibid.
136 GERMANS, IRISH AND
YANKEES Planchon, Les
Vignes américaines . . .,
pp.58–9
137 A SIMPLE HONEST IRISHMAN
ibid., p.70
137 THE EXCELLENT QUALITY
Revue des deux mondes,
15 February 1874, p.941
137 IMMENSE BUILDING
Planchon, Les Vignes
américaines . . ., p.61
137 TO OBSERVE HOUR BY HOUR
ibid.
137 MEANS OF CONTAGION Riley,
Fourth Report . . ., p.64
138 OUR WINGED FEMALE ibid.,
p.65
138 WE FOLLOWED ATTENTIVELY
Planchon, Les Vignes
américaines . . ., p.63
139 FOR THE LADIES "Notes du
voyage," in Morrow, "The

American Impressions . . .,"
op. cit., p.75
140 WAS OFTEN DONE BY
NEGROES ibid.
140 MUCH INFERIOR TO ibid.
141 GIVEN THE AGE-OLD ibid.
141 THE UNHURT Revue des deux
mondes, 15 February 1874,
p.940
142 THOSE WHO RIGHTLY Riley,
Third Report . . ., p.29
142 FIRST WE MIGHT EXPECT
ibid., p.95
142 IT IS THE BATTLE OF LIFE
Revue des deux mondes,
15 February 1874, p.939
143 I DID NOT REGARD Louis
Pasteur, in Phylloxéra
vastatrix: Enquête de
l'Académie des Sciences
pendant les années 1873 à
1878 . . . (Paris, 1879),
pp.24–5
143 A GOOD VIN ORDINAIRE
ibid.
143 THE PREJUDICE WHICH
Planchon, Les Vignes
américaines . . ., p.78
144 AS FOR ORDINARY WINES
ibid.
144 THE INTRODUCTION OF ANY
TRANSPLANT Journal
d'agriculture pratique, 1874,
Vol. 2, pp.75–6
145 HOWEVER UTOPIAN Riley,
Third Report . . ., p.25
145 A QUESTION OF FINDING
ALLIES Revue des deux
mondes, 1 February 1874,
p.564
146 M. PLANCHON "French
Vineyards and Vine Disease,"
The Times, 16 November
1872
146 SINCE NUMEROUS
NEWSPAPER Riley, Sixth
Report . . ., p.55

147 BY ANY MEANS Malvezin,
Lettre à la Chambre de
Commerce de Bordeaux,
pp.35–6
147 LITTLE BIRDS Pétition in
Misc. Reports, France,
Phylloxera 1872–1900, Kew
148 SPEAKING NO FRENCH
Weber, Peasants Into
Frenchmen, p.95
148 THE ENGLISH NAME Journal
d'agriculture pratique, 1878,
Vol. 2, p.734
149 AMERICAN VINES Vialla, Les
Cépages américaines . . .,
pp.4–5
150 WERE THE ONLY TWO
Comptes rendus des travaux
du congrès viticole et
séricicole . . ., p.133
151 AFTER SIX YEARS Compte
rendu du congrès viticole
international tenu à
Montpellier, 1874, p.211
152 IF THE JUDGEMENT OF ibid.,
p.217
152 HARDLY AGREEABLE ibid.,
p.218
153 THE WINE WHICH Compte
rendu du congrès viticole
international tenu à
Montpellier, 1874, p.219
152 IT IS IMPOSSIBLE Daniel, La
Question phylloxérique, p.24
154 FOR FOUR YEARS Journal
officiel, 10 February 1877,
p.1045
154 ROOTSTOCK AND SCION
P. Galet, General Viticulture,
p.281
155 GRAFTING, ALTHOUGH
Guyot, Culture of the Vine
and Wine Making, p.7
155 FOR THE PAST THREE YEARS
Journal d'agriculture pratique,
1881, Vol. 2, p.804
156 THE GIRONDE WILL NOT

"Bulletin du phylloxéra dans
le Gironde," No. 2, quoted in
Roudié, Vignobles et
vignerons du Bordelais, p.170
156 ITS LECTURE HALLS Journal
d'agriculture pratique, 1879,
Vol. 1, p.356
157 IN THE MESSAGER DU MIDI
quoted in Pouget, Histoire de
la lutte . . ., p.47
157 THE ATTILA OF FRANCE'S
VINES, ibid., p.16
157 THE NURSERY MUST
Instructions relatives à
l'établissement des pépinières
de vignes américaines (Paris,
1883), p.3
158 IS A QUESTION Planchon,
Résumé pratique . . ., p.31
159 WHAT'S THE USE OF
LOOKING Planchon, "La
Question du phylloxéra en
1876," in Revue des deux
mondes, 15 January 1877,
p.270
160 THE BARBARIANS ibid.
161 FROM SMALL IDEAS Journal
d'agriculture pratique, 1879,
Vol. 2, p.731
161 INNUMERABLE THEORIES
ibid., pp.146–7
162 A PRIMITIVE PÉNIS Balbiani,
Comptes rendus de l'Académie
des Sciences, 4 October 1875
162 THE SOLE AIM OF Riley,
Sixth Report . . ., p.41
163 FROM THE VERY MOMENT
Balbiani, Comptes rendus de
l'Académie des Sciences, Vol.
81, pp.581–8
164 THE MYRIADS OF SUCKING
Planchon, in Revue des deux
mondes, 15 January 1877,
p.267
164 MINUTE GALLERY Boiteau,
Comptes rendus de l'Académie

des Sciences, Vol. 82, pp.155–7

164 A LITTLE TORTOISE Planchon, Les Moeurs du phylloxéra, p.xviii

165 INDEFINITE ibid., p.3

165 OTHER THAN ON THIS UNIQUE Cornu, Études sur le phylloxéra vastatrix, p.11

166 WITH ALL THE DEFERENCE Congrès International . . . d'Anvers 1885. Rapport de M. Planchon, p.2

167 A LARVA – WHICH Comptes rendus de l'Académie des Sciences, 1877, Vol. 85, p.507

167 A LOCAL ACCIDENT ibid.

168 DEBATE OVER THE Weber, Peasants Into Frenchmen, p.244

168 THE ATTACHMENT TO Loubère, Radicalism in Mediterranean France, pp.112–13

169 TO SURVIVE A VIGNERON Laurent, Les Vignerons de la Côte d'Or . . ., p.352

170 IN VILLAGES WHERE Barral, La Lutte contre le phylloxéra, p.59

170 A WIDOW WHO HAD ibid.

171 THE EFFORTS OF MY Journal officiel de la république française, March 1876, p.1611, in Misc. Reports, France, Phylloxera 1872–1900, Kew

172 THEY WHO KNOW quoted in Pouget, Histoire de la lutte contre le phylloxéra, p.20

172 STRAWBERRY-LOVER de Bompart, La Délivrance de la vigne, pp.7–9

172 RESISTANT AMERICAN Le Congrès phylloxérique internationale (Lausanne, 1877), pp.17–25

173 THE REPRESENTATIVES Planchon to Colonial Secretary, Cape of Good Hope, in Hahn, Report on some questions connected with viticulture at the Cape, p.20

174 STAYED ON THE quoted in Pouget, Histoire de la lutte contre le phylloxéra, p.41

174 CHOICE COLLECTION OF "Report of the United States Commissioners [to the Paris Universal Exposition of 1878] . . .," p.345

174 THE NEW LAW FORBIDS Journal d'agriculture pratique, 1878, Vol. 2, p.218

174 FRENCH WINE Laliman, "Étude sur les divers phylloxéra," p.148

175 WHICH WILL DECIDE, BY VIRTUE ibid.

176 ALL THE AGRICULTURAL Paul Leroy-Beaulieu, L'Économiste français, August 1879, p.155

178 I REGRETFULLY Comptes rendus de l'Académie des Sciences, 1877, Vol. 85, p.509

178 PROPRIETORS TO BE Journal d'agriculture pratique, 1878, Vol. 1, p.435

178 A KIND OF APATHY ibid.

179 INSPECTION MUST BE CARRIED OUT Journal d'agriculture pratique, 1879, Vol. 2, p.435

185 AUSTRALIAN WINE IS Wine and Spirit News, 22 July 1882, pp.57–9

186 EXACTLY THAT OF letter in Wine Trade Review, 15 January 1881, p.11

186 A SPARSE POPULATION David, Phylloxera Vastatrix, the Grape Vine Destroyer, pp.22–3

186 GENETIC RESEARCHERS "Sex and the Single Female Phylloxera," in *La Trobe University Bulletin*, November 2001

187 NOTHING NOW APPEARS Viala, *American Vines*, p.7

189 SAVE FROM DESTRUCTION Carosso, *The California Wine Industry*, p.118

189 THE CAUSE OF THIS ibid., p.114

191 MUST IT BE Millardet, *Notes sur les vignes américaines*, p.73

192 THE PHYLLOXERA CRISIS *Le Temps*, 19 August 1878

193 SECRET LIQUID Jouet, *Historique de la reconstitution de Saint Bénézet*, p.74

193 THE FIRST MERCHANT Fitz-James, *Les Congrès viticoles*, p.58

194 GRANDS CRUS AND SMALL Fitz-James, *Le Congrès de Bordeaux*, p.xvi

194 WE CONTINUE TO BE *Le Temps*, 19 August 1878, p.5

194 YOU FORGET IN CONDEMNING *Le Temps*, 26 August 1878, p.3

195 FIGHT MUST ibid.

196 YOUR VINE IS DEAD *Journal d'agriculture pratique*, 1878, Vol. 2, p.733

196 ARAMON AND CARIGANES ibid.

197 BURGUNDY HOPES TO *Revue des deux mondes*, January 1877, p.277

198 THE PROPRIETORS *Wine and Spirit News*, 27 May 1882, p.391

199 IF WE ARE GOING TO DEFEND quoted in Pijassou, *Un Grand vignoble . . .*, p.753

199 THE VINES SO TREATED *Wine and Spirit News*, 11 March 1882, p.194

200 MASCULINE SEX Garrier, *Vignerons du Beaujolais . . .*, p.126

201 SEAL OFF ALL ACCESS Côte d'Or departmental records, quoted in Laurent, *Les Vignerons de la Côte d'Or . . .*, pp.329‒32. Also C. Ladrey, "Rapport sur l'invasion du phylloxéra dans le département de la Côte d'Or" (Paris, 1878), pp.4‒9

201 AFTER THE GRAPE-HARVEST Laurent, *Les Vignerons de la Côte d'Or . . .*, p.335

201 PHYLLOXERA WAS THE FRIEND ibid., p.336

202 AMONG THE DEMONSTRATORS ibid., p.337

203 SCIENCE WILL EMERGE "Rapport du Commission Supérieure du Phylloxéra, 1881," in *Journal d'agriculture pratique*, 1882, Vol. 1, p.130

204 BRUTUS www.circustef.com

204 TWENTY THOUSAND Fitz-James, *Les Congrès viticoles*, p.46

205 LITTLE NURSERY ibid., p.49

205 CONTINUED TO SPEAK ibid., p.33

205 NORTHERN CHINA Lavalée, *Les Vignes Asiatiques et le phylloxéra*, p.8

205 BUT WHAT SORT OF WINE *Journal d'agriculture pratique*, 1879, Vol.1, p.780

206 IN SPITE OF THE RESPECT Planchon, *Encore les vignes de soudan*, p.2

206 THE MAHARAJAH OF *Wine Trade Review*, 15 September 1881, p.432

207 NOTWITHSTANDING THE
ibid., p.15
207 INDUCE SMALL WINE-
FARMERS Hahn, Report on
some questions connected
with viticulture at the Cape,
p.11
208 FRENCH WINE COMING
"Congrès Internationale
Phylloxérique" (Bordeaux,
1882), p.39
209 LED THE GIGANTIC Le
Temps, 12 October 1881, p.3
209 IF YOU WILL LET ME ibid.
209 THE EXTREMELY ANIMATED
Hahn, Report on some
questions connected with
viticulture at the Cape, p.16
210 WIPED OUT EVERY SUMMER
Journal d'agriculture pratique,
1881, Vol 2, p.618
210 WOOD MERCHANTS Fitz-
James, Le Congrès de
Bordeaux, p.xvi
210 BATTAILLON CARRÉ Fitz-
James, Les Congrès viticoles,
p.51
210 THAT HUSSARD DE LA MORT
Fitz-James, Le Congrès de
Bordeaux, p.xi
210 HAPPY AND CONFIDENT
ibid., p.v
211 THE HARVESTERS Journal
d'agriculture pratique, 1881,
Vol. 2, p.804
211 THE IMMUNITY OF
AMERICAN Hahn, Report on
some questions connected
with viticulture at the Cape,
p.17
211 WHAT RICH WINE Barral, La
Lutte contre le phylloxéra,
p.133
212 THE DENIAL OF ALL Fitz-
James, Les Congrès viticoles,
p.51
211 ONE OF THESE WAS FIRMLY

Journal d'agriculture pratique,
1881, Vol. 2, p.618
212 MANY AMERICANISTS Fitz-
James, Les Congrès viticoles,
p.50
213 OBSERVATION 1 "Congrès
Internationale Phylloxérique"
(Bordeaux, 1882), p.314
213 THESE TASTINGS ibid., p.356
214 WINES WHICH Hahn, Report
on some questions connected
with viticulture at the Cape,
p.17
214 TO OUR TASTE Thistleton-
Dyer, International Phylloxera
Congress of Bordeaux, p.6
215 INVENTORS FROM ALL PARTS
"Rapport du Commission
Supérieure du Phylloxéra," in
Journal d'agriculture pratique,
1884, Vol. 1, p.267
216 THE BLESSING OF THE SEEDS
ibid.
216 EVERY TIME IT WAS quoted
in Pouget, Histoire de la lutte
contre le phylloxéra, p.41
217 THE ENORMOUS PROFITS
ibid., p.88
217 THE PHYLLOXERA – THIS
TIME quoted in Garrier,
Vignerons du Beaujolais . . .,
p.130
217 THE DEMAND FOR ibid., p.87
218 PRETENDED ANTI-
PHYLLOXERA Le Petit
Bourguignon, 13 December
1885, quoted in Laurent, Les
Vignerons, p.347
218 JUST TWO OR THREE Pouget,
Histoire de la lutte contre le
phylloxéra, p.89
219 1887 WILL MARK A DATE
ibid., p.69
222 WHOLE VILLAGES H. Hitier,
"Éloge de Pierre Viala,"
Revue de viticulture,
17 March 1938, p.197

224 ONE OR SEVERAL Viala, *Une Mission viticole en Amérique*, p.ix

225 HOPE OF FINDING ibid., p.xii

227 ALTHOUGH MANY SPOKE quoted in Pouget, *Histoire de la lutte . . .*, p.78

233 EVERY NEWSPAPER *Journal d'agriculture pratique*, 1890, Vol 2, pp.219, 325, 385

234 THE ADMINISTRATION *L'Illustration*, 18 August 1894, p.135

235 ABNORMAL REPRODUCTION ibid.

235 WILL THIS FOUNTAIN ibid.

237 DOGMAS OF RECONSTITUTION Lucien Daniel, "The Crisis in the Vineyard," *The Times*, 21 July 1908, p.16

238 "MÉDOC" OUT OF WATER Pijassou, *Un Grand vignoble . . .*, p.819

238 CHÂTEAU DES GÉANTS Roudié, *Vignobles et vignerons du Bordelais*, p.208

238 IT IS SUPERFLUOUS *Le Progrès agricole et viticole*, 21 June 1908

239 SOCIALISTIC LAND "Secrets of the Wine Trade," *Daily Telegraph*, 28 August 1908, p.4

240 THE CRIME OF FALSE *Ridley's Wine and Spirit Trade Circular*, 8 June 1905, p.460

240 MILLIONS OF GALLONS *Ridley's Wine and Spirit Trade Circular*, 8 March 1905, p.202

240 IN GERMANY WHERE ibid., p.201

243 IT WAS NOT THE PHYLLOXERA Lachiver, *Vins, vignes et vignerons*, p.464

243 PRESENTED A WHITE

"Compte rendu des travaux du service du phylloxéra," reported in *Journal d'agriculture pratique*, 1900, Vol. 2, p.715

245 VIVE LE VIN NATUREL! Garrier, *Histoire sociale . . .*, pp.356–60

245 THE WINE CRISIS Lachiver, *Vins, vignes et vignerons*, p.467

250 THIS INNOVATOR Fitz-James, *Les Congrès viticoles*, p.6

251 ACCUSED OF quoted in Pouget, *Histoire de la lutte . . .*, p.16

251 TRIUMVIRATE Laliman, *Notice chronologique . . .*, p.16

251 ALL OF OUR ibid., p.43

251 ALL HE ASKS OF FRANCE Ordish, *The Great Wine Blight*, pp.219–20

252 M. LALIMAN'S CLAIM ibid.

252 THE NATIONAL REWARD ibid.

252 M. LALIMAN WAS THE FIRST *Journal d'agriculture pratique*, 1897, Vol. 2, p.813

253 RURAL DEMOCRACY *Journal d'agriculture pratique*, 1894, Vol. 2, p.849

253 DISCOVER, UNDERSTAND ibid.

257 CABBAGE-HEADS *American Journal for Enology and Viticulture*, Vol. 34, No. 2, 1983, p.86

258 THE VERY RARE entomology.ucdavis.edu/ faculty/granett/phypage.htm (May 2000)

259 I HAVE NOW COLLECTED entomology.ucdavis.edu/ faculty/granett/downie.htm

259 THE SIZE OF A WALK-IN M. Wood, "A Possible Preventive

for Phylloxera," *Agricultural Research*, December 1998
259 SEXUAL FORMS *Phylloxera Phlyer Newsletter* (UC Davis), January 1997
260 RESEARCHERS CAN www.ars.usda.gov/is/AR/archive/dec98
261 IN FIVE TO TEN YEARS *Grape Research News*, Vol. 7, No. 2, 1996
263 THE GRAPEVINE IS A COMPOUND PLANT *New Scientist*, 17 July 1997
263 IT SHOULD IN THEORY www.slowfood.it
263 STUDY OF THE MECHANISMS press release, Institute for Horticultural Development, Knoxfield, Victoria, 13 February 1998
264 TO DEVELOP AN INDUSTRY *New York Times*, 13 July 2001

265 TERRE ET VIN www.tvbtvm.online.fr
265 BURGUNDY AND BORDEAUX *Wine Spectator*, 31 May 2001
267 THE REAL DISASTER Le Gris, *Dyonisos crucifié*, p.149
268 I HAVE TASTED PRE- *Wine Spectator*, 31 August 2000
269 AS IS THE CASE www.chateaulosboldos.com
269 THE LIVING RELIC www.champagne-bollinger.fr/us/les_vins/vignes_franc.html
270 THE FIRST GUST OF WIND Newsgroup: fr.rec.boissons.vins, 1 September 2002
270 INCREDIBLE, ITS TASTE ibid.
270 THREE VERY OLD www.wineandco.com/fr
271 LA MÉMOIRE DE LA VIGNE J.-C. Bouvier, www.gluiras.com/vag/vag80/oenologie
271 OF CONSISTENT COLOR ibid.

BIBLIOGRAPHY

The literature on phylloxera is immense. The ampelographer Pierre Galet in his businesslike *Les Maladies et les parasites de la vigne* lists almost two thousand titles, most of them published in the years when the aphid reigned in the vineyards of Europe from 1870 to 1900. As the plague retreated, so did the tide of words; when it returned in the late twentieth century to stalk a globalized wine industry, so did the exchange of information. The internet is full of news of the *puceron*.

The malady of the vine was a mass phenomenon. As well as the scientific discourse played out in the proceedings of the Académie des Sciences and the journals of learned societies there was an outpouring of polemical journalism and pamphleteering wherever the vine was grown and its products consumed. These published contemporary accounts, many hundreds of them, preserved in libraries at Montpellier, Davis, California, and at Kew, provide the core sources for this story. The astonishing website of the antiquarian wine-book specialist Eberhard Buehler of Toronto (bookdaemon-.com) pointed to many more routes of inquiry.

Journals such as the *Messager agricole du Midi* provided frontline reports of the disaster's unfolding, while the pages of the *Journal d'agriculture pratique* gave a detailed running commentary throughout thirty years of the crisis.

Manuscript sources include the letters of J.-E. Planchon to the Hookers, father and son, held at Kew, and copies of the botanist's papers retrieved from the Montpellier School of Entomology and the Planchon family in the 1950s by the late American agricultural historian Dwight D. Morrow, now preserved in his papers at Davis.

French departmental archives have been mined by several generations of historians in search of the political and social effects of the plague. Especially useful were the works of Gilbert Garrier (le Beaujolais), Robert Laurent (la Côte d'Or), René Pijassou (le Médoc), Philippe Roudié (le Bordelais), A.R.H. Baker (Loir-et-Cher) and Iain Stevenson (l'Hérault). M. Galet's many viticultural works were a constant guide on technical matters, as were the entries by Dr. Richard Smart in the very accessible *Oxford Companion to Wine* (ed. Jancis Robinson). Without reading at the outset *The Great Wine Blight*, published thirty years ago by the English agronomist George Ordish, the whole subject would have been totally impenetrable.

CONTEMPORARY PUBLICATIONS

Académie des Sciences, *Commission du phylloxéra. Séance du 3 Décembre 1874* (Paris, 1875)

Académie des Sciences, *Observations sur le phylloxéra et sur les parasitaires de la vigne* (Paris, 1881–84)

Anonymous, *La Vigne, le phylloxéra et les canaux d'irrigation* (Avignon, 1874)

Balbiani, Édouard Gerard, *Observations sur la reproduction du phylloxéra du chêne* (Paris, 1874)

Barral, Jean-Augustin, *La Lutte contre le phylloxéra* (Paris, 1883)

Berlandier, Jean-Louis, *Journey to Mexico during the Years 1826–1834* (Texas State Historical Society, 1981)

Boiteau, P., *Guide pratique du viticulteur pour la destruction du phylloxéra* (Libourne, 1877)

Bordeaux, *Compte rendu général du Congrès International phylloxérique . . .* (Bordeaux, 1882)

Bouschet, Henri, *Moyens de transformer promptement par les vignes américaines les vignobles menacés par le phylloxéra* (Montpellier, 1874)

Buchanan, Robert, *The Culture of the Grape and Wine-Making* (Cincinnati, 1856 and 1865)

Busby, James, *Journal of a Recent Visit to the Principal Vineyards of Spain and France* (London, 1839)

Bush, Isidor, *Les Vignes américaines. Catalogue illustré et descriptif . . . Bush & Son and Meissner* (trans. Planchon and Bazille, Montpellier, 1876)

California, *Report of the Committee on Phylloxera, Vine Pests and Diseases of the Vine* (Sacramento, 1881)

Catalogue de Bushberg (trans. Montpellier, 1875)

Cazalis-Allut, Louis, *Mémoires sur l'agriculture . . . et l'oenologie* (Montpellier, 1848)

Cazalis-Allut, L., *Mélanges de viticulture, d'oenologie et d'agriculture* (Montpellier, 1859)

Charmet, M., *Le Phylloxéra: Ses ravages et les moyens de le reconnaître et de le détruire* (Lyon, 1873)

Compagnie des Chemins de Fer de Paris à Lyon et à la Méditerranée, *Rapport sur les travaux effectués par ce service pendant la campagne de 1882* (Marseille, 1883)

Cornu, M., *Le Phylloxéra, la nouvelle maladie de la vigne* (Paris, 1874)

Cornu, M., *Études sur le phylloxéra vastatrix* (Paris, 1879)

Crolas, Ferdinand, *Guide du vigneron pour l'emploi du sulfure de carbone contre le phylloxéra* (Lyon, 1884)

Daniel, Lucien, *La Question phylloxérique, le greffage et la crise viticole . . .* (Bordeaux, 1908)

Darwin, Charles, *The Movements and Habits of Climbing Plants* (London, 1875)

Darwin, C., *The Variation of Animals and Plants under Domestication* (2nd edition, New York, 1883)

David, G., *Phylloxera Vastatrix, the Grape Vine Destroyer* (Brisbane, 1878)

de Bompart, Amélie, *La Délivrance de la vigne et la découverte du trombidion dévorateur du phylloxéra* (Paris, 1879)

Demole-Ador, *Le Congrès phylloxérique international* (Lausanne, 1877)

Donnadieu, A.L., *Sur Quelques points controversés de l'histoire du phylloxéra* (Paris, 1887)

Dunn, Malcolm, *The Horticultural Handbook* (Edinburgh, 1895)

Faucon, Louis, *Mémoire sur la maladie de la vigne et sur son traitement par le procédé de la submersion* (Montpellier, 1874)

Fitz-James, Marguerite Augusta Marie Löwenhjelm, *Le Congrès de Bordeaux* (Nîmes, 1882)

Fitz-James, M., *Grande culture de la vigne américaine 1881–1883* (Paris, 1884)

Fitz-James, M., *Les Congrès viticoles* (Paris, 1889)

Flagg, William J., *Three Seasons in European Vineyards; treating of vine-culture; vine disease and its cure; wine-making and wines, red and white; wine-drinking, as affecting health and morals* (New York, 1869)

Flagg, W.J., *Handbook of the Sulfur Cure* (New York, 1870)

Foëx, Gustave Louis Émile, *Rapport à M. le directeur de l'École d'Agriculture de Montpellier* (Montpellier, 1879)

Foëx, G., *Ampélographie américaine: Description des variétés les plus intéressantes de vignes américaines* (Montpellier, 1885)

Foëx, G., *Cours complet de viticulture* (1st edition, Montpellier, 1886)

Gachassin-Lafite, Léon, *Études sur le phylloxéra vastatrix dans le Gironde . . .* (Bordeaux, 1873)

Gachassin-Lafite, L., *Conservation des vignobles par la rhizoplastie* (Bordeaux, 1874)

Gastine, G., *Traitement des vignes phylloxérées: Emploi du sulfure de carbone contre le phylloxéra* (Paris, 1884)

Gervais, Prosper, *Adaptation et reconstitution en terrains calcaires* (Montpellier, 1896)

Girard, Maurice, *Le Phylloxéra de la vigne, son organisation, ses moeurs et choix des procédés de destruction* (Paris, 1874)

Guyot, Jules, *Culture of the Vine and Wine Making* (trans., Melbourne, 1865)

Guyot, J., "La Viticulture et ses produits," in *Exposition Universelle de 1867. Rapports du Jury International . . .*, Vol. 12 (Paris, 1868)

Guyot, J., *Études sur les vignobles de France*, Vol. 3 (Paris, 1868)

Hahn, Paul, *Report on some questions connected with viticulture at the Cape. Report by Roland Trimen . . . the delegate appointed to represent this colony at the International Phylloxera Congress recently assembled at Bordeaux* (Cape Town, 1882)

Hérault, Department of the, *Commission Départementale de l'Hérault de la maladie de la vigne caractérisée par le phylloxéra* (Montpellier, 1873)

Hérault Commission Départementale . . ., *Expériences faites à Las Sorres près Montpellier . . .* (Montpellier, 1877)

Hilgard, Eugene W., *The Phylloxera or Grapevine Louse, and the Remedies for its Ravages* (Sacramento, 1880)

Husmann, George, *The Cultivation of the Native Grape* (New York, 1866)

Jouet, M.L., *Historique de la reconstitution de Saint Bénézet* (Nîmes, 1884)

Ladrey, C., *Station oenologique de la Côte-d'Or. Le phylloxéra. Histoire de la nouvelle maladie de la vigne* (Paris, 1875)

de Lafitte, Prosper, *Quatre ans de luttes pour nos vignes et nos vins de France* (Paris, 1883)

Laliman, Léopold, *Réformes viticoles. Cépages indigènes de l'amérique* (Paris, 1860)

Laliman, L., *"Étude sur les divers phylloxéra,"* reprinted from *Messager agricole du Midi* (Montpellier, 1871)

Laliman, L., *Études sur les divers travaux phylloxériques et les vignes américaines . . .* (Paris, 1879)

Laliman, L., *Notice chronologique sur l'origine des vignes américaines résistant au phylloxéra . . .* (Bordeaux, 1889)

Laval, Henri, *De la Maladie de la vigne dans le département de Vaucluse* (Montpellier, 1869)

Lavalée, Alphonse, *Les Vignes Asiatiques et le phylloxéra* (Paris, 1878)

Lenoir, B.-A., *Un Traité de la culture de la vigne et de la vinification* (Paris, 1828)

Lichtenstein, Jules, *Histoire du phylloxéra* (Montpellier, 1879)

Lyon, *Comptes rendus des travaux du congrès viticole et séricicole* ... (Lyon, 1873)

Mach, Edmund, "Die Phylloxera Vastatrix in Frankreich," *Annalen der Oenologie*, Vol. 3 (1872), pp.462–85

Malvezin, Théophile, *Lettre à la Chambre de Commerce de Bordeaux sur le phylloxéra de la vigne* (Bordeaux, 1874)

Marion, Antoine, *The Phylloxera: A Short Treatise on the Vine Destroyer with a Report upon the Bi-sulphide of Carbon* (Paris, 1880)

Mayet, Valéry, *Les Insectes de la vigne* (Paris, 1890)

Millardet, Alexis, *La Question des vignes américaines au point de vue théorique et pratique* (Bordeaux, 1877)

Millardet, A., *Études sur quelques espèces de vignes sauvages de l'Amérique du Nord* (Bordeaux, 1879)

Millardet, A., *Notes sur les vignes américaines* (Bordeaux, 1881)

Millardet, A., *Histoire des principales variétés et espèces de vignes d'origine américaine* (Bordeaux, 1885)

Montpellier, *Compte rendu du congrès viticole international* ... (Montpellier, 1875)

Mouillefert, Pierre, *Le Phylloxéra. Moyens proposés pour le combattre* (Paris, 1876)

Mouillefert, P., *Le Phylloxéra. Résumé des résultats obtenus en 1876 à la station viticole de Cognac* (Paris, 1877)

Odart, Alexandre Pierre, Count, *Ampélographie, ou traité des cépages les plus estimés dans tous les vignobles de quelque renom* (Paris, 1845)

Planchon, Jules-Émile, and Saintpierre, Camille, *Premières expériences sur la destruction du puceron de la vigne* (Montpellier, 1868)

Planchon, J.-E., and Lichtenstein, J., "Notes entomologiques sur le phylloxéra vastatrix," appendix to L. Vialla, *Phylloxéra et la nouvelle maladie de la vigne* (Montpellier, 1869)

Planchon, J.-E., and Lichtenstein, J., *Le Phylloxéra en Angleterre et en Irlande* (Montpellier, 1871)

Planchon, J.-E., *Le Phylloxéra. Faits acquis et revue bibliographique (1868–1871)* (Montpellier, 1872)

Planchon, J.-E., and Lichtenstein, J., *Le Phylloxéra (de 1854 à 1873). Résumé pratique et scientifique* (Montpellier, 1873)

Planchon, J.-E., *Les Vignes américaines, leur culture, leur résistance au phylloxéra et leur avenir en Europe* (Montpellier, 1875)

Planchon, J.-E., *Les Moeurs du phylloxéra de la vigne. Résumé biologique* (Paris, 1877)

Planchon, J.-E., *Encore les vignes de soudan* (Lyon, 1881)

Ponsot, Mme., veuve, *Les Vignes américaines* (Paris, 1890)

Rafinesque, Constantine, *American Manual of the Grape Vines and the Art of Making Wine* (Philadelphia, 1830)

Riley, Charles Valentine, "The Grape-Leaf Gall-Louse – Phylloxera," in "Third annual report on the noxious, beneficial and other insects of the State of Missouri"; "Sixth annual report of the State Board of Agriculture, Jefferson City, 1871"; also seventh, eighth and ninth annual reports (Jefferson City, 1872–76)

Riley, C.V., *The Locust Plague in the United States: Being more particularly a treatise on the Rocky Mountain Locust or so-called Grasshopper . . . with practical recommendations for its destruction* (Chicago, 1876)

Rohart, François Ferdinand, *État de la question du phylloxéra* (Paris, 1875)

Sahut, Félix, *Les Vignes américaines* (Montpellier, 1885)

Sahut, F., *Un Épisode rétrospectif à propos de la découverte du phylloxéra* (Montpellier, 1900)

Signoret, Victor, "Phylloxéra vastatrix . . . cause prétendue de la maladie actuelle de la vigne," in *Société Entomologique de France, Annales* (Paris, 1870)

Thistleton-Dyer, W.T., *International Phylloxera Congress of Bordeaux* (Kew, 1882)

Trimoulet, A.-H., *Rapport sur la maladie nouvelle de la vigne* (Bordeaux, 1869)

Trimoulet, A.-H., *3e Mémoire sur la maladie de la vigne* (Bordeaux, 1874)

Trimoulet, A.-H., *5e Mémoire sur la maladie de la vigne* . . .
remèdes préconisés par les phylloxéristes . . . (Bordeaux,
1875)
Viala, Pierre, *Une Mission viticole en Amérique* (Paris, 1889)
Viala, P., *American Vines: Their Adaptation, Culture, Grafting
and Propagation* (translation, Melbourne, 1901)
Vialla, Louis, *Le Phylloxéra et la nouvelle maladie de la vigne.
Étude comprenant le rapport de la Commission nommée par
la Société des Agriculteurs de France pour étudier la
nouvelle maladie de la vigne* (Montpellier, 1869)
Vialla, L., *Les Cépages américains dans le département de
l'Hérault pendant l'année 1876* (Montpellier, 1877)
Victoria, Australia, "Papers respecting phylloxera vastatrix or
new vine scourge" (Melbourne, 1873)
Wilder, Marshall P., "Report of the United States
Commissioners [to the Paris Universal Exposition, 1867]
upon the Culture and Products of the Vine" (Washington,
D.C., 1868)

MODERN STUDIES

Baker, A.R.H., *Fraternity among the French Peasantry:
Sociability and Voluntary Associations in the Loire Valley,
1815–1914* (Cambridge, 1999)
Boehm, W., *The Phylloxera Fight* (Adelaide, 1996)
Browne, Janet, *Charles Darwin: The Power of Place* (London,
2002)
Carosso, Vincent, *The California Wine Industry: A Study of
the Formative Years* (Berkeley, 1951)
Galet, Pierre, *Les Maladies et les parasites de la vigne*, Vol. 2
(Montpellier, 1977)
Galet, P., *Grape Varieties and Rootstock Varieties* (trans.,
Chaintré, 1998)
Galet, P., *General Viticulture* (trans., Chaintré, 2000)
Garrier, Gilbert, *Vignerons du Beaujolais au siècle dernier*
(Roanne, 1985)

Garrier, G., *Le Phylloxéra: Une guerre de trente ans, 1870–1900* (Paris, 1989)

Garrier, G., *Histoire sociale et culturelle du vin* (Paris, 1995)

Lachiver, Marcel, *Vins, vignes et vignerons* (Paris, 1988)

Laurent, Robert, *Les Vignerons de la Côte d'Or au XIX siècle* (2 vols, Paris, 1957)

Le Gris, Michel, *Dyonisos crucifié* (Paris, 1999)

Loubère, Léo, *Radicalism in Mediterranean France* (New York, 1974)

Loubère, L., *A History of Wine in France and Italy in the Nineteenth Century* (New York, 1978)

Maillet, Pierre, *Contribution à l'étude de la biologie du phylloxéra* (Paris, 1957)

Morrow, Dwight D., "The American Impressions of a French Botanist," *Agricultural History*, Vol. 34 (Washington, D.C., 1960)

Morrow, D. D., *The Phylloxera Story* (unpublished manuscript, UC Davis)

Mullins, Michael G. (ed.), *Biology of the Grapevine* (Cambridge, 1992)

Ordish, George, *The Great Wine Blight* (London, 1972)

Paul, Harry W., *Reinventing the Vine: Science, Vine and Wine in Modern France* (Cambridge, 1997)

Pijassou, René, *Un Grand vignoble de qualité: Le Médoc* (Paris, 1980)

Pinney, Thomas, *A History of Wine in America from the Beginnings to Prohibition* (Berkeley, 1989)

Pouget, Roger, *Histoire de la lutte contre le phylloxéra de la vigne en France (1868–1895)* (Paris, 1990)

Roudié, Philippe, *Vignobles et vignerons du Bordelais* (Bordeaux, 1988)

Savage, H. and E.J., *André and François André Michaux* (Charlottesville, 1986)

Stevenson, Iain, "Viticulture and Society in the Hérault (France) during the Phylloxera Crisis, 1862–1907," in *Journal of Historical Geography*, Vol. 6, 1980

Weber, Eugen, *Peasants Into Frenchmen: The Modernization
of Rural France* (Stanford, 1976)

NEWSPAPERS AND JOURNALS

FRANCE
Comptes rendus de l'Académie des Sciences
L'Économiste français
L'Illustration
Journal d'agriculture pratique
Journal de l'agriculture
Messager agricole du Midi
Messager du Midi
Le Monde illustré
Revue de viticulture
Revue des deux mondes
Le Temps
La Vigne américaine

GREAT BRITAIN
Daily Telegraph
The Gardener's Chronicle and Agricultural Gazette
Journal of the Horticultural Society of London
Ridley's Wine and Spirit Trade Circular
St. James's Gazette
The Times
Wine and Spirit News
Wine Trade Review

UNITED STATES
Cozzens' Wine Press
Phylloxera Phlyer Newsletter
Scientific American
Wayward Tendrils
Wine Spectator

ILLUSTRATIONS

In the late 1860s grape-vines in south-east France began to wither and die. Not until an investigative committee dug up an apparently healthy vine and found its roots encrusted with tiny yellow aphids was the villain revealed. (*British Library, Colindale*)

By the mid-1870s the phylloxera's life-cycle was more or less understood, and France's vine-growers had been mobilized in "study and vigilance committees" armed with recognition charts of the invader. (*British Library*)

Attempts to eradicate the plague by uprooting and burning infected vines were too little and too late. Windlasses and pulleys or an outsize crowbar called "the goat," were used to extract vines from the ground. Later this wheeled crane was devised. (*British Library*)

The Montpellier botanist Jules-Émile Planchon realized that the aphid was the cause of the disaster, and that it had somehow come from America. (*La Bibliothèque de l'École Nationale Superieure Agronomique de Montpellier, ensaminra*)

Charles Valentine Riley, state entomologist of Missouri, found the insect on the roots of vines in the United States, and came to France in 1871 to prove the connection. (*University of California, Riverside*)

Professor Georges Balbiani of the École de France, whose winter-egg theory led to another diversion in finding effective means to combat the plague. (*British Library*)

21, rue Longue in the little town of Roquemaure in the Gard. Planchon traced the first outbreak of the disease to the walled vineyard behind this modest establishment. (*Author photograph*)

307

A map of infected vineyards around Dijon shows the "oil-spot" pattern of the malady's spread. (*ensam-inra*)

Inundating vineyards to drown the underground insect worked on the plains, but not on hillsides. (*British Library*)

Carbon bisulphide was known to be an effective insecticide. The problem was how to get the explosive chemical deep enough into the ground. Giant cast-iron syringes were devised, and the P-L-M railway company recruited a small army of technicians to wield them. (*British Library*)

Hundreds of bizarre proposals were advanced in bids to win the prize of 300,000 gold francs for an effective and practical remedy, including this "death-chamber." (*Shields Library, UC Davis*)

Desperate vine-growers grasped at whatever opportunist manufacturers might proclaim as sovereign remedies for the plague. Mr. Davis of Boston came to France in 1875 to promote the merits of his "compound insect destroyer." (*ensam-inra*)

Potassium carbonate was found to be an effective insecticide. It needed large quantities of water to get it into the ground; only grand proprietors could afford the necessary steam pumps and pipework. (*British Library*)

In 1873 Planchon was sent to the United States to further research the life-cycle of the insect and the potential of American vines to resist its attacks. (*Shields Library, UC Davis*)

Montpellier's School of Agriculture became the intellectual citadel of the *américainistes*. (*ensam-inra*)

As the crisis deepened there was a scramble to get hold of American vines. J.P. Berckmann, nurseryman of Augusta, Georgia, circulated this pamphlet around southern France in May 1874. (*Shields Library, UC Davis*)

After the first wave of transatlantic shipments, French nurserymen began to satisfy the demand for American vines with home-grown specimens. Léo Laliman was not going to let allegations of having imported the aphid in the first place interfere with a colossal commercial opportunity. (*Royal Botanic Gardens Library*)

The winter-egg theory produced several proposed remedies,

including this chain-mail glove with which *vignerons* were supposed to scrape vine-stems clean of infected bark.
(*British Library*)

Professor Jean-Baptiste Dumas, chairman from 1870 of the government-appointed Central Commission on the Phylloxera. (*La Bibliothèque centrale du Muséum national d'histoire naturelle*)

Alexis Millardet, the Bordeaux botanist and geneticist who established which American species were truly resistant to the phylloxera. (*La Bibliothèque centrale du Muséum national d'histoire naturelle*)

Professor Maxime Cornu of the Paris Museum of Natural History, secretary of the Phylloxera Commission, who conducted important research on how the aphid attacked vine roots. (*La Bibliothèque centrale du Muséum national d'histoire naturelle*)

Pierre Viala, professor of viticulture at Montpellier, was sent on a government mission to Texas in 1887 to find phylloxera-resistant vines that would grow in chalky soil (*ensam-inra*).

Camille Saintpierre, pioneer in the fight against the aphid and later director of the Montpellier School of Agriculture. (*Author photograph*)

The memorial to Jules Planchon, in the Place Planchon opposite Montpellier's railway station. (*Author photograph*)

The sensuous memorial to Gustave Foëx at La Gaillarde. (*Author photograph*)

The experimental vines at La Gaillarde, flanked now by ultra-modern research facilities. (*Author photograph*)

INDEX